Trusting Measurement Results

In the Chemical and
Process Industries

D0843745

Also Available from ASQ:

Quality Assurance for the Chemical and Process Industries: A Manual of Good Practices, Second Edition
ASQ Chemical and Process Industries Division

ISO 9000 Guidelines for the Chemical and Process Industries,
Second Edition
ASQ Chemical and Process Industries Division

Specifications for the Chemical and Process Industries: A Manual for Development and Use
ASQ Chemical and Process Industries Division

A Laborartory Quality Handbook of Best Practices and Relevant Regulations
Donald C. Singer, editor

An Introduction to Design of Experiments: A Simplified Approach
Larry B. Barrentine

Improving Performance through Statistical Thinking
ASQ Statistics Division

Root Cause Analysis: Simplified Tools and Techniques
Bjørn Andersen and Tom Fagerhaug

Quality Problem Solving
Gerald F. Smith

To request a complimentary catalog of ASQ Quality Press publications, call 800-248-1946, or visit our online bookstore at http://qualitypress.asq.org .

Trusting Measurement Results

In the Chemical and Process Industries

ASQ Chemical and Process Industries Division
Chemical Interest Committee

ASQ Quality Press
Milwaukee, Wisconsin

Trusting Measurement Results in the Chemical and Process Industries
ASQ Chemical and Process Industries Division, Chemical Interest Committee

Library of Congress Cataloging-in-Publication Data

Trusting measurement results in the chemical and process industries / by Chemical and Process Industries Division, ASQ, Chemical Interest Committee.
    p. cm.
Includes bibliographical references and index.
ISBN 0-87389-503-7 (alk. paper)
    1. Chemical industry—Quality control. I. American Society for Quality. Chemical Interest Committee.

TP150.Q34 T78 2001
660'.068'5—dc21
    2001022322

10 9 8 7 6 5 4 3 2 1

ISBN 0-87389-509-6

Acquisitions Editor: Annemieke Koudstaal
Production Administrator: Gretchen Trautman
Special Marketing Representative: David Luth

ASQ Mission: The American Society for Quality advances individual and organizational performance excellence worldwide by providing opportunities for learning, quality improvement, and knowledge exchange.

Attention: Bookstores, Wholesalers, Schools and Corporations: ASQ Quality Press books, videotapes, audiotapes, and software are available at quantity discounts with bulk purchases for business, educational, or instructional use. For information, please contact ASQ Quality Press at 800-248-1946, or write to ASQ Quality Press, P.O. Box 3005, Milwaukee, WI 53201-3005.

To place orders or to request a free copy of the ASQ Quality Press Publications Catalog, including ASQ membership information, call 800-248-1946. Visit our web site at www.asq.org or http://qualitypress.asq.org .

Printed in the United States of America

 Printed on acid-free paper

American Society for Quality

Quality Press
P.O. Box 3005
Milwaukee, Wisconsin 53201-3005
Call toll free 800-248-1946
Fax 414-272-1734
www.asq.org
http://qualitypress.asq.org
http://standardsgroup.asq.org
E-mail: authors@asq.org

Table of Contents

Preface

The Chemical and Process Industries Division (CPID) of the American Society for Quality (ASQ) is committed to quality improvement throughout the chemical and process industries (C&PI). Since 1984 the Chemical Interest Committee (CIC) has worked to reach consensus on good quality practices for our industries and documented them in a series of books. The first book, published in 1987 by ASQC Quality Press, was *Quality Assurance for the Chemical and Process Industries—A Manual of Good Practices*. It acquired the nickname the "Little Red Book" and has been a best seller for Quality Press. The second edition of the "Little Red Book" was released in 1999.

The ISO 9000 standards for quality systems were first published in 1987. The CIC saw a need to link our concepts of good practices to the new ISO 9000 standard. In 1992 the CIC published ANSI/ASQC Q90 *ISO 9000 Guidelines for Use by the Chemical and Process Industries*, another best seller for Quality Press. The CIC updated this book and title following the release of the 1994 version of ISO 9000 as *ISO 9000 Guidelines for the Chemical and Process Industries*, Second Edition.

In 1996 the publication of *Specifications for the Chemical and Process Industries—A Manual for Development and Use* documented for the first time recommended practices for specifications. This was another important step toward our goal of clarifying issues and stating consensuses on essential practices for the C&PI.

Now, this new volume on trusting measurement results provides clarification and consensus on the essentials of measurement systems in C&PI. Trusted measurement results are essential to trusted business decisions.

Acknowledgements

The CIC members who participated in this book's writing and editing are:

Georgia Kay Carter	Jack Herman
Ken Chatto	Norman Knowlden
D. C. Cobb	William Ochs
Jeffrey Dann	Frank Sinibaldi
David Files	Ben Wood
Terry Geary	Priscilla Zawislak

The support of the following companies is hereby acknowledged:

Kenneth A. Chatto, Consultant

Boehringer Ingelheim Chemicals, Inc.

Eastman Kodak Company

Hercules Incorporated

OSRAM *SYLVANIA* Products Inc.

Quantum Chemical Company

John T. Herman Consulting

Frank Sinibaldi Consulting

Witco Corporation

Global Business Connection, LLC

AlliedSignal, Inc

Philip Stein, ASQ Fellow and Past Chair of the ASQ Measurement Quality Division, assisted in the review and content.

1

Introduction

1.1 PAST AND PRESENT

A century ago, Lord Kelvin issued his famous dictum, "When you can measure what you are speaking about, and express it in numbers, you know something about it; but when you cannot measure it, your knowledge is of a meager and unsatisfactory kind."[1, 5] Lord Kelvin's dictum highlights the importance of measurement values and measurement results for quantifying the differences among things, so that communication between people and decisions about those things can be more effective, consistent, and correct.

Traditionally, measurement results came from on-site laboratories. Today, a measurement laboratory is often a subcontractor or is otherwise geographically separated from where the sample is taken and from the location of the user of the measurement results. Measurement systems have also been impacted by computer technology, causing a tendency to accept the reported measurement results as trustworthy. Both of these situations make it harder and more costly to verify the measurement results. Yet, the pressure is on us to make data driven decisions, for example, management by fact, lean manufacturing, Six Sigma, and SPC. We are becoming a society where people are abdicating their responsibility for making decisions to "the numbers made me do it!" way of life. Every number that we receive or generate begs the questions: How should the number be used? What is the full meaning of the number?

This book is your tool for finding out if the numbers you receive are trustworthy and for you to build trust into the numbers you report.

1.2 STRUCTURE OF THIS BOOK

In our daily work and personal lives, we are bombarded with numbers from measurement systems that we can only imagine, and some that we never dreamed existed. We are asked to trust all of them. Our common sense says that while many are trustworthy, there are more than a few instances where trusting the numbers is a big mistake. Every number raises many questions, and most of the time we do not have the answers needed to decide whether to trust the number. We have tried to include all of the components of a measurement system and describe the aspects of each to help answer your questions.

Chapter 1 sets the stage for the rest of the book by relating real stories and discussing the concepts that form the foundation for the book. Chapter 2 describes the components of a measurement system and their impact on our ability to trust the measurement values and measurement results. Chapter 3 describes the use of measurement values and measurement results to make key decisions about the measurement system and the material or process variable measured. Chapter 4 takes a look into the future of measurement technology and its application. Chapter 5 contains appendices with more in-depth presentations on key issues.

1.3 DO YOU RECOGNIZE ANY OF THESE REAL WORLD STORIES?

If you see yourself in these stories or have your own measurement story, this book is for you.

1.3.1 I'm Stuck, What Do I Do Now?

"Frank, I'm stuck, what do I do now?" asked Rita, lead operator on a high-density polyethylene reactor. "We just got the lab result from the in-line grab sample, and it indicates we're not only outside our SPC limits, but out of spec as well. The continuous melt indexer has been trending downward, but certainly not this far."

Frank walked over to Rita's station and peered over her shoulder to look at the reading on the display screen of the distributive control system. "Tell Bill to grab another sample and get it over to the lab ASAP for a retest. What's changed in the last eight hours?" Frank walked over to the logbook and read the journal entries of the previous 24 hours while Rita changed menus and reviewed secondary data listed in the distributive control system.

"Looks like there was a catalyst batch change three hours ago," Frank said in a distant voice.

His mind was already racing ahead to the decision point and his stomach was feeling the stress from a dilemma with no straightforward solution. With the reactor pumping out 45,000 lb/h, he knew a lot of cost was in the balance. To make matters worse, yesterday he told Marketing that the reactor was running smoothly and he could increase the rates to handle orders from a new customer. In addition, inventory levels were near record lows. Although this made the executive staff and stockholders happy, it meant higher stress levels for Manufacturing and the Supply-chain Management Team because unplanned outages for mechanical or quality problems could not be tolerated, especially now.

Which instrument value should he trust? Reactor set points would be adjusted based on which test was believed. The continuous melt indexer had a tendency to drift over time. He also had experience with lab "fliers," test results so different than expected that the operators knew not to trust them. Retest often confirmed their suspicions. However, this lab result was right on the border of believability, and the catalyst change lent it some credibility. Two hours could elapse before the retest would come back. The new result often added to the confusion rather than clarifying it. In the meantime, a full 185,000-lb lot would be produced and the next one would start.

"Rita, average the two results (original and retest) and make a step change to the modifier level. Also, call the lab and ask them to turn the sample as quick as they can. Get someone down to the continuous melt indexer and recalibrate it. Also, inform the Material Management Team about this lot; tell them to keep an eye on the melt index results because the lot may have to be downgraded." Frank was already thinking about all the possible scenarios. He knew the set point change would not be exactly correct but that further adjustment would be necessary. He also knew that set point changes took time to fully take effect. With a bed change of six hours, it may take longer before the next set point change would be known. It was going to be a long, tense night.

1.3.2 All Is Not Well at the Paper Mill

The area had been involved with SPC training and its application to a refiner, which basically takes a water slush of market pulp at a given consistency and refines it to develop fiber strength by lowering freeness. Consistency is a weight concentration measure and freeness is a water release measure. The higher the freeness, the faster pulp slurry will drain water. Freeness was tested every three hours from a sample taken at the refiner.

At 7:15 PM, the first sample was taken by the analyst and tested for freeness. The number was reported back to the kraft mill about 30 minutes later. The reported number was very high and outside of the upper control limit for the freeness control chart. About that time the shift team began its on-the-floor problem solving training with George. As luck would have it, a problem did exist—what to do about this out-of-control point? Until this current freeness number, the control charts were in control with no problems indicated for the last two days. George facilitated the team into listing possible courses of action for the problem freeness number:

• Take another sample and retest;

• Chalk it up to a bad test and wait until the next test time (~10 PM); or

• Check for equipment problems.

After checking out the equipment and finding nothing wrong, the team decided to ask for a retest. The analyst returned to the area and resampled the refiner. At 8:30 PM, the lab reported the retest number. This result was 20 units higher than the 7:15 PM test. With this confirmation, the team began to brainstorm possible causes of high freeness. They quickly decided that the most likely cause was a bad sample. The decision was that they would collect another sample and take it to the lab for testing. It was now 9:30 PM, and the third number was obtained. You guessed it, same story—still high and out-of-control.

The team was divided. Half of them felt that something was definitely wrong while the others felt they were victims of a measurement problem. Several choices were discussed:

• Call their boss at home;

• Ask the shift area manager to get a freeness test run by the main lab; or

• Wait until the regular 10 PM sample is tested.

They finally reached the consensus that they had a measurement problem and not a process problem. They waited. Around 11 PM, the regular test result was received. This time it was about 50 units lower, but still above the upper control limit. The team spent about an hour rechecking the production process and trying to decide on what action to take.

To make a long story short, the team finally decided (at 1 AM) that the cause of the high out-of-control points was due to an analyst error in running the freeness test. Their decision was presented in the form of a bet with George: "We will do nothing and the freeness will come back into control with the first sample taken after the shift change (at 6 AM)." George lost the bet when subsequent investigation by the main lab team revealed that the high results were in fact due to a consistent analyst error. The test method was not being followed and resulted in the consistently high freeness results. Retraining of the second shift control lab analysts was planned.

1.3.3 Variable Weight Loss Sounds Like a Diet Story

The production operators' testing area was audited. One measurement was percent moisture using a weight loss instrument. Operators reported that they were having problems with the moisture content of the product. The measurement results varied greatly from sample to sample and the customers experienced variations in performance.

The procedure called for placing 10 grams of product on the instrument. The auditor observed the actual measurement and discussed the steps in the measurement with several people who carried it out. They were careful to be exact about the 10 grams. Some would take off material if they went over 10 grams, others would add material slowly so they did not go over the 10 grams.

The auditor also examined the calibration records and asked about the procedure the calibration company used. There was no information on the calibration company's procedures. The records showed that the instrument was calibrated at 100- and 300-gram points. There was no determination of the sensitivity of the scale. The procedure required the instrument to be able to detect a change of 0.0015 grams, but there was no calibration information to support that the instrument could detect this change. The act of being precise about the amount of material placed in the instrument causes erroneous measurement results. The material is losing moisture during the act of putting a precise amount on the instrument. Three factors were making the measurement of moisture a wasted activity:

1. The instrument was not calibrated or monitored for its ability to detect the required change of 0.0015 grams;

2. The material was losing moisture because of the excessive handling during additions to moisture balance; and

3. There was a general lack of knowledge and training in the correct practices for determining moisture content using the instrument.

Could the recipients of these moisture content numbers make any proper use of the data? No! If they had asked these questions, they might have found the problems, not used the measurement results, and requested corrective action. Was the material losing water during handling? Was the instrument calibrated to prove that the required sensitivity exists? Is the uncertainty of the measurement system known?

1.3.4 Everyone Was Happy

The Acme Company processes about a million pounds of tungsten ore, scrap, and sludge a month. The price of tungsten is $10.22 per pound. Tungsten content is determined by X-ray fluorescence. Joe Moe is the engineer in charge of the X-ray laboratory. In calibrating his instrument he uses a NIST standard reference material which is reported to contain 67.22 ± 0.35% (2 standard deviations) tungsten.

Joe always reported the results of his testing to two decimal places, until one day Nan Cann in the accounting department decided that since a difference of 0.005 percent in tungsten analysis could equal thousands of dollars a year, Joe should report tungsten content to three decimal places.

Joe objected, saying that the precision of the measurement does not justify reporting three decimal places. He explained that if each sample were run about a dozen times then the average could be reported to that third decimal place. "No, no," said Nan, "that would be too expensive and time consuming. Just give me the three decimal places that I asked for." Joe decided that there was only one thing to do. He would report to three decimal places, but the third decimal would always be a zero. And everyone was happy.

1.3.5 Statistical Thinking Is Important in Research

Researchers often say that statistics is important to production where there are thousands and thousands of data points, but not of much use in product or process research were data sets are relatively small. On the contrary, statistical practices are most powerful in preventing the wrong conclusion being drawn from too small a data set. An excellent argument for using statistics in research is the following story of an actual research project. It is the tale of a golf ball taken "in the tail."

A midwestern company that had made its fortune in rubber technology, especially the bonding of rubber to other materials and the development of useful mechanical devices, decided that their knowledge of rubber should make it easy to develop the perfect golf ball. A team was formed and soon the researchers were producing a golf ball with spectacular characteristics. Long distance was the first aim of the researchers, because everyone in golf knows that distance sells. Within six months the research team was ready to show management their progress. They brought "Iron Mike," the mechanical golf ball-hitting machine, to the country club and management watched as the research golf balls beat every other golf ball on the market by 15 to 20 percent in the distance trial. Management was ecstatic. Golf ball manufacturers were brought in to view the trials. Again the research balls outperformed every other golf ball, a point not missed by the manufacturers. A bidding war broke out, contracts were signed, and management began to anticipate the great profits. Instead, deliveries were late and the balls were rejected. The manufacturer that bought the research balls was suing for nonsupply. What began as one of the company's most successful research projects was becoming a financial disaster and public relations embarrassment.

How could this happen? An industrial statistician was hired to investigate the project failure. He found that the researcher began the project using single position production stations. Each ball was handmade from scratch. The experiments were designed to screen different materials, winding schemes, and covers. The designs did not consider interactions. Three shifts of 20 technicians were kept busy screening different combinations of materials, winding schemes, and covers. The output of each technician was tested by measuring the distance "Iron Mike" hit the balls. The longest-hitting ball was chosen as the best combination. Once the combination was chosen, all the technicians began producing the same ball using the same material, winding scheme, and cover. Each day's output was tested. Balls that did not meet the performance criteria were thrown away. The failures were attributed to defects in workmanship that researchers believed would be overcome when the balls were manufactured mechanically. The researcher kept only 2 percent of the balls produced during the entire project. If the researcher had taken the time to plot a histogram of all measurement results, he would have seen all the balls produced came from the one distribution. On average, the research balls were no better than any other golf ball. It was only the extreme top 2 percent of the production that met the performance criteria. To make matters worse, when the statistician examined the data from the original experiments, he discovered there never had been a super golf ball. The entire company had been knocked into a hazard by a golf ball "in the tail."

1.4 PURPOSE OF MEASUREMENT SYSTEMS AND MEASUREMENT RESULTS

Measurement results provide objective data that differentiate options under consideration so that a decision or choice can be made among the options. *Defining the decision or choice to be made is the starting point for choosing or developing a measurement system.* Making the decision based on the numbers generated by the measurement system is the end of the decision cycle. *The measurement system starts with taking the sample to be measured from the population and ends with delivering the measurement result needed to make the decision.* Those responsible for making the decision are the customers of the measurement system.

The type of decision governs the measurement system that will be used. For certain decisions, the measurements can be "coarse" (miles), other decisions require very fine (millimeters) highly differentiating measurements. The smallest difference required for the decision defines the divisions of the unit of measurement (UOM) scale that the measurement system must differentiate. Every activity in the measurement system should be chosen for its contribution to the delivery of a trustworthy measurement result that will support a correct decision.

1.5 WHAT IS A MEASUREMENT?

The International Vocabulary of Measurement (VIM)[6] defines a measurement as "a set of operations having the object of determining a value of a quantity." The value determining operation is a comparison, thus for this book we define measurement method as "a qualitative or quantitative comparison of two things at a time resulting in the assignment of measurement values." All measurement methods are comparisons—whether we compare with our senses (eyes, hands, nose, ears, and taste), compare a chemical reaction that creates something visible, or a sophisticated electronic instrument compares things using electrons—comparison generates the measurement value.

The most basic measurement method is qualitative comparison of a pair of objects (or samples) to each other: determining which is longer, which is darker, or which is heavier. Even this basic method can have major variation issues and not give a clear measurement result. The binary comparison system can be used to define a rank-ordered set of objects. For example, five chemical samples can be compared for whiteness, one pair at a time, until the most white to least white sample is differentiated and ranked. Then, decisions can be made on the basis of the ranking.

A measurement method can also be a quantitative comparison to a series of marks on a scale. For example, scales for length and weight are defined by fixed points derived from physical standards, plus a method of interpolating between pairs of fixed points (often linear, but not always). Once a scale is established, measurement tools such as spring scales and graduated cylinders can be made and calibrated, and then used independent of the physical standards. By using measurement scales, comparisons may be made at different times, in different locations, and by different people. The fixed points are explicit, quantitative, and accepted by those using them. The comparison to a measurement scale is generally an interpolation between two adjacent markings, one on either side of the sample. Some technicians have been trained to choose and report the nearest scale marking without interpolation, but in fact statistical studies at the National Bureau of Standards have shown that most observers can estimate to within 0.15 of a division (2 sigma limit), and that with training, 0.05 of a division may be discerned.

Digital scales introduce an entirely different set of reading errors. The instrument still has a scale, but the display prevents reporting anything but the marked values, and electronics decide how to report a value that in fact lies between two scale points. Often, digital instruments don't display enough resolution and useful information is lost.

Measurement values reported by some measurement methods offer a choice of several different units of measure depending on the differences required to make the decision. For example, we could report feet, inches, half inches, or smaller with a ruler. Depending on the decision, we may need to report only feet; or we may need to report to one thirty-second of an inch. In both cases, the measurement method is the same, but the required precision in the measurement value is different.

There is variability in all comparisons. This is in addition to any variability in the chemical or physical reaction employed to detect the analyte. The amount of variability that can be tolerated by the measurement method or by the decision influences the choice of measurement methods. Quantifying the variability helps determine the confidence needed in the measurement result. If we are to make decisions using the measurement results, the variability needs to be small enough so that we can identify real differences.

1.6 MEASUREMENT SYSTEM AND PRODUCTION PROCESS RELATIONSHIP

Perhaps the most common and important type of measurements made in the C&PI are for material composition. We need to know material composition for safety, environment, regulatory compliance, cost control, pricing, process control, and customer satisfaction. Measurement is our only means of knowing what elements and compounds are in the materials we produce and use. Composition measurements are also our most challenging because most are inferred measurements (see Section 1.7).

Composition measurements only tell us what we have measured, not what we have not measured. If a measurement method does not detect a component we will not know it is present. Yet we cannot afford to measure every material for everything known in the world. *Selection and design of*

composition measurements requires thorough knowledge of the raw mate-
rials used and of the production process that created the material being
measured. Once designed, all changes in raw material supply and production process require evaluation of their impact on the measurement system.

Has a new material component been introduced that will not be measured by the current measurement system? If the production process is not
capable, no amount of product measurement will make the product good. *A*
measurement system cannot make bad material good. Nor can statistics
make bad data good. Doing statistical analysis on measurement results
from samples that are contaminated, deteriorated, not representative, taken
at the wrong time, wrong location, and so on, is a total waste.

This book presents the role of each component in a measurement system and the variability associated with it. In light of this variability a business can balance its resource expenditure among raw material quality,
production process control, and measurement capability. Controlling raw
materials through supplier quality programs ensures consistent raw materials that make production process design and control easier, as well as reducing measurement costs for raw materials and products. Spending resources
on product design, process design, and production process control so that
there is the capability to deliver conforming product reduces the resources
needed for measurement systems.

If raw materials are bought without regard to quality, if product design
and process design just happen, and if process control is an afterthought,
then no amount of measurement system resources is enough. If the raw
material control and production process design and control prevent impurities, their measurement should require fewer resources. If impurities can
enter from several sources and by-products can vary with raw material
source and process conditions, greater measurement resources will be
needed to cover all possibilities. *Pay a little up front or pay much more later.*

1.7 REACTION-DEPENDENT
MEASUREMENT SYSTEMS

It should be easier to trust a measurement result when the comparison is
between two objects we can see with our eyes rather than when the measurement result is inferred. That is, we trust the weight of an object more than the
amount of acid in a sample inferred by the amount of base used to neutralize
it, or the presence of a carboxylic acid group inferred by infrared spec-
troscopy.[5,6] In the C&PI, there are many measurement values that are inferred
based on chemical and/or physical reactions that are more or less specific for
the analyte of interest. It is inferred that the amount (measurement value) of
reagent used is only consumed by the desired or expected reaction.

Additionally, the reactions are generally not reversible, so they cannot be repeated with the exact same sample, even if the sample could be isolated in its original form. It is quite common that two reactions are involved in getting a measurement value. These are the key reasons that a representative sample is so important in C&PI and why it is necessary to report many measurement results as averages of several measurement values. While measurement method chemical and physical reaction issues contribute randomly to variation in the measurement system, their impact can be consistent, causing a repeatable bias.

The existence of a measurement method in a compendium of methods does not mean that its chemical or physical reaction is without interferences that are significant in the current usage. All measurement methods should be thoroughly evaluated for the specific analyte being measured and for interferences in the sample environment under consideration. Ruggedness testing is especially applicable to reaction-dependent measurement methods. (See Section 2.2.8, "The Need for Ruggedness Tests.") There are notable cases where a compendium method involving a double inference was shown to be in error by as much as 6 percent because of interferences. Single inference methods that have no interferences should be used where possible.

The scientific considerations of material sampling, analytical chemistry, and physical chemistry are essential to trusting measurement results in C&PI. This book does not address the analytical and physical chemistry considerations involved in C&PI measurements. There are many books and articles on these subjects.[1-4] Most material composition measurement systems are reaction dependent. So, in addition to understanding the contents of Chapter 2 and Chapter 3 of this book from a measurement science perspective, the reader may need to include the issues associated with single or multiple inferred measurement values and the calculations that are an integral part of those measurement systems.

1.8 QUALITY SYSTEM'S ROLE

The contents of this book require the existence of a functioning quality system encompassing the raw material, production, and measurement areas of the company. Because the first impact of raw material changes is on the measurement system requirements, the supplier quality component of the quality system is important to maintaining measurement system capability. Each measurement system should be a documented part of the quality system. *A functioning quality system provides the infrastructure needed to generate trustworthy measurement results.*

1.9 REFERENCES

1. Steven M. Stigler, *The History of Statistics—The Measurement of Uncertainty before 1900* (Cambridge, MA: Belnap Press of Harvard University, 1986).
2. W. Funk, V. Dammann, and G. Donnevert, *Quality Assurance in Analytical Chemistry* (New York: VCH, 1995).
3. I. R. Juniper, "Method Validation: An Essential Element in Quality Assurance," in *Quality Assurance and TQM for Analytical Laboratories*, ed. M. Parkany (Royal Society of Chemistry, 1995).
4. Peter C. Meier and Richard E. Zund, *Statistical Methods in Analytical Chemistry* (NY: John Wiley & Sons, 1993).
5. Charles Seife, *Zero—The Biography of a Dangerous Idea* (New York: Penguin Putnam, 2000).
6. International Organization for Standardization, *The International Vocabulary of Measurement*, Second edition (Geneva, Switzerland: International Organization for Standardization, 1993).

2

Is This a Trustworthy Number?

That is the *big question* to which the impact of variation holds the answer. Measurement results are the output of measurement systems that have multiple components of variation. We must look for and quantify variation from the moment the population to be sampled is identified to the point when the measurement result is reported. These activities define a measurement system. This question has a user's (customer's) view:

A. Can we, as users, trust the numbers being given to us by the lab and the inline instruments?

and a producer's view:

B. Can we generate numbers that all of us can trust?

These two questions have a common-thread variation:

1. Do we know the sources of variability in the material being measured and in the measurement system generating the measurement result? To use a measurement result we need to be confident in the generation of that measurement result.

2. Do we understand the impact of the sources of variability on the usefulness of the measurement result? To deliver a trustworthy measurement result, we need to know how to correctly handle variation during generation of that measurement result.

2.1 WHERE DO I START?

Start with the intended use of the measurement result, which dictates the required measurement system. This part addresses the generation of measurement values and measurement results. Chapter 3 addresses key uses and associated decisions.

2.2 THE KEYS TO TRUSTING A MEASUREMENT VALUE OR MEASUREMENT RESULT

In order to trust and use a measurement value or measurement result:

- The decision criteria must be defined;

- The measurement system must be suitable for its intended use (the decision);

- The measurement method must be appropriate for the purpose;

- The sample must represent the population; and

- The format for reporting measurement value(s) and measurement results must be appropriate for the purpose.

The measurement system encompasses everything from taking the sample through generating the measurement value to reporting the measurement result. The measurement system must be under controlled conditions throughout the generation of the measurement value. Every activity, method, person, material, and piece of equipment in the measurement system affects the reported measurement result. Furthermore, every action, method, person, material, and piece of equipment has variability that introduces uncertainty into the measurement results.

2.2.1 The Decision and Unit of Measure Must Be Defined

A clear definition of the decision to be made is required to choose a proper measurement system and to determine if a measurement result is fit for use in making the decision. Is the required decision between broad categories or between small differences? We need to know the smallest difference in the characteristic/property of interest that is required to make the decision. This difference defines the precision of the unit of measure that the measurement system must be capable of differentiating with the required confidence. For example, to decide if a material contains sodium, a flame identification test could be used. To quantify that the amount of sodium that is in the material to 0.000001 gram would require an atomic absorption measurement.

2.2.2 The Measurement Method Must Be Suitable for Its Intended Use

The measurement method:

- *Must be able to distinguish the differences that have physical or chemical significance to the situation under consideration.*

- *Must have variability that can detect differences that have physical or chemical significance to the situation under consideration.*

- *Must have the precision and accuracy needed for the decision.*

Evaluating the measurement system should involve determination of the significant physical and chemical differences by people who are knowledgeable of the situation under consideration and the measurement method, and who have the capability of determining the statistical significance of the measurements. How much of a difference must be known before the measurement of that difference can be designed. Once variation of measurement values, amount of difference, and the desired risks are known, the number of measurement values needed to make the decision can be calculated. Developing or selecting the best measurement systems usually involves the cooperation of analytical chemists, statisticians, owners of the situation under consideration (chemist/chemical engineer), and any others with appropriate knowledge.

Determining the suitability of the measurement method for intended use typically involves:

- Accuracy/measurement uncertainty

- Bias

- Precision

- Limit of detection (LOD)

- Limit of quantification (LOQ)

- Selectivity

- Linearity

- Range of interest

- Ruggedness

These are often referred to as validation of the measurement method.[1, 3, 4]

2.2.3 The Sample Must Represent the Population

A sample is expected to be a representative part, or single item, taken from a population. Statisticians and others ordinarily want to generalize from a small body of data (sample) to a larger body of data (population). The term *representative sample* is often used to connote a sample taken from a population (universe) that is expected to exhibit the same properties as the population (universe).

The definition of the options for the decision and the characteristics or properties of interest determine the sampling requirements. Improper samples can lead to misinformation, erroneous decisions, errors, recalls, and so on. *A measurement result can be no better than the sample(s) measured.*

Therefore, to be of decision-making value, the measurement must be made on a sample that represents the population. Although the following discussion of populations and sampling is focused on material characteristics and properties, sampling is by no means confined to materials. Measurement of production process parameters, such as rpm, temperature, pressure, speed, flow rate, and so on also involves samples and populations. For example, when measuring rpm, the population could consist of all the rpm values that could have been measured during the time to run a batch production process, and a sample could be a set of measurements made during a period of time during the batch production process.

A representative sample possesses the same properties in the same proportions as the larger mass or bulk material—therefore, it is very important to ensure that the sample represents the population from which it is taken. It is equally important that neither the population (bulk material) nor the sample become contaminated during the sampling process. There are additional considerations:

- Safety—the taking of the sample must neither endanger the person taking the sample nor people in nearby areas.

- Environmental—the environment should neither be harmed nor affected by the sampling process.

- Economical—the sampling process must consider the balance between the needs for precision and economics of sampling and measurement costs.

2.2.3.1 Sampling Definitions

Sampling is the process of removing a selected portion of material from a larger mass for the purpose of making inferences regarding the larger mass. *Sampling* is a *process*—it has input (population), an operating procedure, and output (sample), each of which must be subject to control. *Sampling is the* first *and* most important *process within the measurement system.*

Knowledge that helps determine the sampling process:

- Subsequent use of the material;

- Historical information on past supplies of the material; and

- Equipment suitable for sampling such material.

An operating procedure for sampling must be written and followed. The *sample* is the initial output of sampling. In process industries, the sample is subjected to observation and/or measurement, often analytical, and the resulting outputs are used to make inferences regarding the mass from which the sample was removed. For any sample, the population is only the immediate larger mass from which the sample was taken.

2.2.3.2 Sampling in the Chemical and Process Industries

Routine inspection and measuring in the C&PI typically includes incoming inspection, in-process inline or offline measuring, and finished goods inspection. Sampling decisions for these inspections and measurements should provide the information needed for suitable control of material variability for usage in the next process or stage. Homogeneity implies smaller variability.

For incoming material, repetitive sampling (several samples from the same population) establishes information on the variability (homogeneity) of material for its usage in the next production process stage or by the customer. To demonstrate such control, multiple samples should be taken from multiple locations in the bulk container (rail car, barrel, bale, and so on). Once homogeneity is established for a material, by means of control charts and/or R&R (repeatability and reproducibility) studies, reduced sampling may be considered. (See Section 3.3.2.2, "Repeatability and Reproducibility (R&R) Studies.")

In-process sampling must continue at each stage to assure that the production process remains in control. Inline or online samplers are often used. The user is responsible for demonstrating that this equipment takes *representative samples* from the production process. The next production process stage, any in-process storage between stages, and packaging for the consumer should be considered when designing the sampling process. Final product sampling should consider the nature of the bulk source, packaging material, package size, and the intended manner of usage by the customer.

The following illustrates and explains the population in sampling of incoming material for inspection:

The population is the totality of material under consideration. Since there are several points at which the material is under consideration, there can be several populations for any given material, each of which might be sampled.

The manufacturer (supplier) has an in-process population, which may end up in several bulk storage containers (silos, tanks, and so on). Each storage container is a sample of the in-process population and is, itself, a storage population which may be sampled by packaging (cardboard containers, drums, carboys, and so on) into one or more container units for shipment. Each container unit is then a population. Any one of these populations may be referred to as manufacturer's bulk.

The customer population is the totality of material purchased from the manufacturer for use at the customer's facility. The customer population is generally made up of one or more shipments from the supplier. Each incoming shipment is itself a population and also a sample of the manufacturer's bulk. Within this incoming shipment population there may be individual containers, each being a sample of that incoming shipment and a population from which container samples can be taken. In some circumstances further sub-sampling occurs, forming further populations, until a specimen is taken for measurement from the final sub-sample. Figure 2.1 illustrates such population/sample levels.

Each stage is a *sample* of the previous stage and a *population* for the subsequent stage. As far as the sample is concerned, the population is only the immediate larger mass from which the sample was taken.

The final sub-sample must be less than or equal to the quantity required in the next production process stage, and it should be made as uniform as

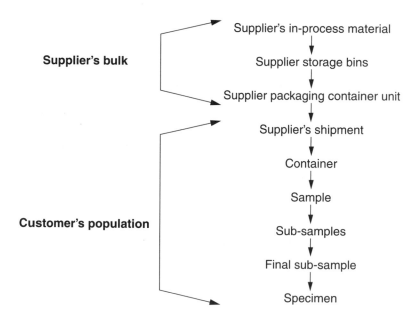

Figure 2.1   Population/sample levels.

feasible by mixing, stirring, pulverizing, and so on. When multiple readings on one specimen from the final sub-sample are attainable, they will provide an estimate of the variability of the measurement method and an average estimate of the level of the property of interest in that particular specimen. When the specimen is destroyed, our estimates have the additional variability from specimen to specimen however small it has been made.

Measurement results on multiple specimens from the final sub-sample (now the population), analyzed in similar fashion, will provide an estimate of the variability between specimens, which is also the variability within the final sub-sample, and an improved estimate of the level of content of the final sub-sample. Increasing the number of specimens analyzed will improve the confidence in these estimates, but will not necessarily reduce the estimate of variability. The variability estimated within the final sub-sample must be no greater than the variability limit required in the subsequent usage stage to demonstrate acceptable homogeneity of that final sub-sample population. The total variability at this stage includes measurement variability and mixing variability.

Multiple samples from the final or previous sub-samples (the populations), carried through to analysis in similar manner, will provide an estimate of the level of content of that population, as well as an estimate of the total variability within that sub-sample. If the total variability estimate at each stage is no greater than the required variability limit of the subsequent usage stage, then sub-samples and mixing for uniformity may not be required. Once variability of the container population is demonstrated, it may be sufficient to consider only random specimens from throughout the incoming container.

Samples from multiple containers will provide an estimate of the level of content of the shipment (now the population) and an estimate of the total variability within that shipment. If the estimated variability is no greater than that required for the subsequent production process stage, and it has been established by R&R (repeatability and reproducibility) studies (see Section 3.3.2.3) or by control charts, then it might be sufficient to consider random specimens from throughout subsequent shipments to demonstrate acceptable representation and homogeneity.

An extended history of acceptable shipments, over time, in terms of level of content and total variability between those shipments, may warrant "approval" of the supplier's shipping procedures and justify lessening the number of samples (specimens) taken on each shipment. This assumes the supplier maintains control of his production process, storage, and shipping procedures. Control charts should be maintained on incoming shipments. Any deviation from control requires repeating the procedures to again justify "approval."

2.2.3.3 What Is an "Approved" Production Process or Supplier?

Earlier, we said that an "approved" production process of our own or a supplier's might reduce our sampling requirements. If approved, material can be placed in stock or supplied to a just-in-time production line. But what is an "approved" production process? By whom is it "approved?" The user of the material determines "approval." In the C&PI, the user must know how and in what quantities the material is to be used in each operation and how it was produced and delivered to the subsequent point of use; that is, in what quantities, and how usage and delivery quantities coincide with bulk sources. The user determines the allowable variability of properties between and within lots of material delivered plus between and within quantities used at the subsequent point of use. When there is enough data to show that the product conformity and uniformity are within allowable variability, the production process can be "approved." Until satisfactory control of conformity and uniformity between delivery units and between usage units can be demonstrated (possibly through a history of control charting representative samples), a production process cannot be considered "approved" and sampling/measuring of the material must be continued. Even when "approved," sampling for identification measurements must be carried out, together with periodic complete analysis, to demonstrate continued conformance to allowable variances and continued "approval" of the production process. Were all of the above in place between Ford and Firestone for the supply of tires for the Ford Explorer?[8]

2.2.3.4 Representative Samples

One has to know, or assume, something about the mass sampled and what it will be used for in order to determine what might be a *representative sample*. To demonstrate that the production process is controlled, both in terms of property levels and variability, the sample taken must be representative of the production quantity or period of production involved. When sampling is done by inline samplers for inline or offline measurements, it must be ascertained whether the inline sampler is providing continuous sampling over a time interval, that is, an estimated average, or a point-in-time sample, which requires assuming consistency over the time period. Only when sufficient samples have been taken to demonstrate *homogeneity*, or lack thereof, can the sampling be termed *representative*. Homogeneity is not a requirement of a representative sample. *If you have homogeneity you know you have a representative sample but not the reverse.*

To demonstrate that each quantity of a bulk material that will be used in a production process is equivalent, the sample(s) taken must show that the bulk material is homogeneous. After all, each quantity of material used in the production process should be a representative sample of the bulk material.

2.2.3.5 A Homogeneous Sample

Did the mass come from a homogeneous source? The importance of homogeneity depends on how the mass will be used—all at once in its entirety or in smaller portions. If used in smaller portions, will the sample and its findings remain valid over time?

Homogeneity is a matter of degree. A mass is sufficiently homogeneous (uniform) if, whatever quantity is used in the next stage, any quantity of that size taken from the mass is statistically equal in all properties of importance to any other quantity of that size that might be taken from the mass. In other words, the *variability* of all material properties of importance is so small that all usage units can be considered equal. An indirect measure of *homogeneity* is the *variance* between sample usage quantities. The examples below consider different homogeneity requirements and cases. While ASTM E826[5] is written for reference materials, much of it applies to any material.

2.2.3.5.1 A Drug Tablet

For a drug tablet, the definition of homogeneity implies that an analysis on an individual tablet should detect an amount of active ingredient within statistical and medical limitations equal to the labeled dosage. Furthermore, it implies that a sufficient number of individual tablets from the mass (the "lot"), be individually analyzed to statistically conclude that the variability between tablets is sufficiently small that the difference between tablets is of no practical significance. Only when sufficient samples have been taken to demonstrate homogeneity, or lack thereof, can the sampling procedure produce *representative* samples.

2.2.3.5.2 Blending and Representative Samples

In the making of the drug tablet, assume five pounds of active ingredient were used in a one ton mix and blended for a specified time in a commercial blender. This requires the following assumptions:

- That any five-pound quantity of active ingredient is statistically equivalent to any other five-pound quantity from the same supply (homogeneous).

- That all other ingredients going into the mix meet homogeneity requirements for their important properties.

- That the specified time and method for blending assures uniform blending for the required homogeneity, an important requirement when granule size, structure, or surface properties differ between ingredients. Caution—it is not uncommon to have segregation occur due to excessive blending time.

Representative sampling of the blender output should be sufficient to demonstrate the level of homogeneity between tablet size quantities to meet tablet specifications. Tablet sampling can then be minimized.

The required representative sampling is totally dependent on the nature of usage in the next or subsequent stage. Homogeneity required for uniform pills is inversely measured by the variability that can be tolerated at later stages. For a drug tablet, the homogeneity requirement would be very strict. Homogeneity requirements are dictated by the use of the product. Whether dealing with solids, liquids, gases, or combinations, similar definitions for homogeneity can be applied.

2.2.3.5.3 Samples from a Continuous Production Process

With continuous production processes, sampling is often performed inline. Typically, the sample, taken at fixed intervals in time, is assumed to be a representative sample of a mass defined by that time interval and generally assumed to involve a fixed volume based on flow rate. The mass should really be defined from the midpoints in time between consecutive samples. Again, the assumption is that the sample is taken from a homogeneous process flow produced under standard, well-defined production process conditions. These samples could be considered stratified samples, in units of production time or flow, from a much larger mass, from a longer period of time, or possibly over some batch-limiting condition such as a reactor size. The outcome of such sampling, often presented by control charts, demonstrates the homogeneity (or lack thereof) of the larger mass by this representative sampling. The homogeneity must be sufficient to support the needs of subsequent usage.

Often in so-called continuous production processes, there will be interruptions that really may break the production process into several batch (lot) stages. Representative sampling of a batch nature must be considered when interruptions have occurred. Uniformity may be lost and segregation may occur in surge tanks or storage silos. Density separation of liquids or gases and "rat-holing" of solids are common occurrences.

2.2.3.6 Samples of Nonhomogeneous Sources

There are situations when it is known that the source population is nonhomogeneous. Such populations may be stratified in nature, a mixture of material from different sources, a collection of waste material, or non-normally distributed for other reasons. In some situations, such as bulk sampling, the major purpose of sampling may be to estimate the monetary value of the mass by estimating its average level of active ingredient (for example, coal, wood chips, ore). In other cases, the nature of the stratification may be critical (for example, environmental waste). Non-normal distributions will likely require non-normal (exponential, and so on) procedures (for example, ppm air particle sampling, and so on)

2.2.3.7 Sampling Bulk Material

Fluids (gases and liquids)—Are often stored in bulk storage tanks, shipped in railcars or trucks, supplied in pressurized containers, or processed through pipelines. If the material is composed of one component, either a point-in-time sample or a sample from a continuous sampler is likely. Care must be taken to assure that the continuous sampler is not a dead point providing a sample of some past processing. If the fluid is a mixture, care must be exercised to overcome any differences in density or vapor pressure and any resulting segregation.

Emulsions—Pose difficult sampling problems particularly because of settling and lack of homogeneity. The act of mixing may change the nature of the emulsion.

Viscous liquids—Generally require dip sampling. For mixtures of viscous liquids, segregation is to be expected during storage. Density, vapor pressure, and boiling point differences contribute to segregation.

Slurries (partial solutions or a solid/liquid mix)—May involve automated in-process samplers or point-in-time dip sampling. Segregation is to be expected with any hold-up in the processing flow. All storage facilities should be subject to continuous agitation. Repetitive sampling throughout processing to assure variability control for the next stage is required.

Solids and solid mixtures—May be subject to segregation, not only on standing in bulk storage but also on withdrawal from or filling to bulk storage. Any differences in particle size or material density can lead to "pyramiding" on filling to bulk storage, to "rat-holing" on withdrawal, or to "settling" on long-term storage, resulting in a lack of homogeneity when passed to the next production process stage or to packaging. Continuous recirculation may provide a remedy. The sampling of solid bulk bins may be through inline samplers or by manual sampling on withdrawal. Care must be exercised to assure inline samplers represent the current material and not past material held in a dead space. Sampling of solids from containers requires thief sampling. Use a thief that provides a sample from multiple levels of the container.

Bale—Sampling is similar to container sampling, and requires use of a thief and possibly multiple level sampling.

Cylindrically wound (roll)—Final products such as plastics, textiles, paper, and tape are often limited to beginning and end of roll sampling.

2.2.3.8 Random Compared to Representative Samples

Throughout the discussion of a sample being representative, the idea of a random sample is inherent, that is, as random as a sample can be for a given situation. The taking of replicate samples to demonstrate homogeneity, or lack thereof, to assure that the sampling procedure provides representative samples assumes randomness of a realistic level.

Truly random sampling requires that every portion of the population sampled have an equal probability of becoming that sample. (Portion is that quantity input to the next stage of usage.) In general, truly random sampling is both financially and physically impossible.

For any operation where sampling is performed on a time-frequency basis, the definition of random cannot be met. For gases and liquids, sampling ports limit any consideration of true randomness. Situations that limit sampling to beginning-and-end, such as cylindrically wound material, cannot be random. For materials in storage silos, railcars, or other bulk containers, accessibility negates true random sampling.

Nonhomogeneous sources require even greater effort to achieve random sampling. To determine the overall level of an active ingredient or the location of various strata, it is necessary to adhere, as best one can, to the definition of random.

2.2.3.9 Sample Retention—Packaging and Storage

Often the remainder of the sample has to be retained for possible future review. Sample retention has several requirements:

- *The packaging should not change the properties of the material.* In some cases, it may be helpful for the sample package to simulate the effect of the finished product packaging on the material so that effects over time can be measured.

- *Storage conditions should not change the properties of the material.* Storage conditions for the sample may be required to emulate the product storage conditions so that storage effects can be measured. Storing the samples in the same place as the product can do this. When it is not possible to use the same storage as the product, quantification of the difference between the storage areas becomes a further factor in measuring the difference between two measurement results.

2.2.3.10 Labeling Samples

Labels should provide information so that those handling the sample can take the correct actions with the sample and prevent problems. The following information, which should be required on all sample labels, provides the keys to accessing all other information about a sample and maintaining the sample's integrity:

- Material identification—a unique term for the material. This identification should give traceability to the sampling procedure.

- Sample identification—a unique term for the sample that is linked to the record of the specific production process (lab notebook, process location and container, and so on) from which the sample was taken.

- Required safety and regulatory labeling (HMIS, and so on).

- Date and time—year, month, day, and time that a sample was taken.

- Sampler identification—name, initials, or code to identify the person or device that obtained the sample.

Optional:

- Storage requirements—when specific storage is critical to the integrity of the sample, this requirement must be on the label, for example, refrigeration required, protect from freezing.

(Note: This list does not address labeling for transport and the associated regulations.)

2.2.3.11 Samples and Chain of Custody

It is often important in industrial situations to document the taking, transport, storage, and handling of the sample (chain of custody/paper trail). Chain of custody records provide important documentation for disputes or claims and can facilitate root cause analysis.

2.2.4 Preparing a Sample for Measurement

Many measurements are not performed on the raw sample because the property of interest is not measured directly. Sample preparation must be planned and carried out in a controlled manner following appropriate procedures. Any deviations from the procedures should be recorded.

2.2.5 The People Who Make the Measurements

Just as there is variability in other parts of the measurement system, there will be variability in the actions of trained and qualified people who are involved in the measurement system. Bottom line: A quality system must be in place to support trained and qualified people with checks and balances to help them do their job in the real world.

2.2.6 Measurement Values versus Measurement Results

Measurement values are often confused with measurement results. Measurement values are the output of a device or instrument (raw data). Measurement results are derived from a measurement value or series of measurement values. The measurement result can be a simple mathematical manipulation of a measurement value; for example, the conversion of a pychnometer measurement value into a density measurement result, or the calculation of a measurement result from an average of several measurement values. Failure to recognize the difference between a measurement value and a measurement result can have serious consequences.

The following example shows why it is important to know a measurement value from a measurement result.

The manager of a quality control laboratory that worked with active pharmaceutical ingredients had a specification for a blended product. Six samples from each lot were required for analysis by high pressure liquid chromatography (HPLC) to calculate the lot assay. In this case each analysis was a measurement value. The measurement result was an average of the six measurement values. The manager was new in the position and eager to demonstrate her commitment to following strict FDA current Good Manufacturing Practices (cGMP) requirements. While reviewing lot analysis data before release of the product, she noticed several HPLC measurement values were not within the specification limits. The cGMP requirements strictly forbid averaging out-of-specification measurement results with in-specification measurement results in order to obtain a passing measurement result. The supervisor failed the lot and ordered the specification rewritten to require all HPLC measurement values to be within specification. The outcome was catastrophic. Nearly 27 percent of all lots failed the modified specification. It took two years before another manager spotted the problem. He realized that a measurement value had been mistaken for a measurement result. It took several months of data and statistical analysis before the FDA would allow the specification to be corrected.

Note that neither approach is acceptable without taking into account the size of each lot nor the amount of active ingredient used in the next process stage, so that within-lot and measurement variability can be properly excluded from the specification limits.

2.2.7 Reporting Measurement Results

The key to reporting a measurement result is to report the full meaning of the measurement result, not just the number. *A number without its full meaning is not a measurement result.* Full meaning includes the number, its statistic, the uncertainty about it, and the reasons for the uncertainty. The user of the measurement result should know all of this to make an informed decision. In cases of multiple reports of measurement results on the same material from the same production process, the full meaning needs to be sent with the first report and updates to the meaning sent only when something changes.

When a measurement result (Y) is near the limit of detection (LOD) or the limit of quantification (LOQ), one of the following recommendations should be used for reporting (as an example, assume that the LOD is 3 and the LOQ is 10.):

When the Y is less than the LOD, such as 2, report either:

Description	Example
Less than LOD	< 3
None detected (report the LOD value)	None detected (LOD = 3)

When the Y is between LOD and LOQ, such as 7, report either:

Description	Example
Greater than LOD and less than LOQ	> 3 and < 10
LOD value to LOQ value	3 to 10
Approximately Y (report LOQ value)	Approximately 7 (LOQ = 10)
Detected (report the LOD and LOQ values)	Detected (LOD = 3 and LOQ = 10)

2.2.8 The Need for Ruggedness Tests

Analytical laboratories often have a measurement method handed down from a corporate laboratory, a customer, a compendium like USP, ACS, and so on that does not work. Often, it is because a measurement variable has a large influence on the measurement method and tolerances for running the measurement method are neither given nor known. It could be ambient temperature, the flow rate of a gas, the volume of a reagent, the quality of the water (distilled versus de-ionized versus tap), some component of a reagent, and so on.

Ruggedness tests are used to find the measurement system variable(s) for which small variation has quite large influences on the results of the measurement method. These are the variables that must be controlled if the measurement method is to be useful.

Ruggedness tests are recommended during development and improvement studies of a measurement method to determine the measurement system variables that have a large influence on the method. Ruggedness tests require only a small number of measurement values, such as eight for a typical design involving up to seven variables. The purpose of a ruggedness test is to find which measurement method variables strongly influence the measurement method and to determine how closely these variables must be controlled (the tolerances on their variation).

Ruggedness tests are particularly useful in the C&PI when measurement methods singly or multiply infer measurement values and employ chemical and physical reactions. (Note that ruggedness tests do not determine optimum conditions for the measurement method.) Ruggedness testing is described in detail in ASTM E1169, *Standard Guide for Conducting Ruggedness Tests.*[6] Ruggedness tests often use Plackett-Burman design of experiments with two levels per measurement variable.

Ruggedness testing is a valuable preliminary step before interlaboratory comparison (ILC) of measurement systems, as it helps set tolerances for measurement method variables. The ruggedness test should be run within a single laboratory, so only the effects of changes in the measurement method variables from high to low are seen. The high and low values chosen for each variable should be consistent with those seen over time in typical operations.

The design of the ruggedness test requires simultaneous change of the levels of all the variables, and allows the determination of the separated effects of each variable on the measurement results. For an N=8 ruggedness test, up to seven measurement variables are set at two levels so as to span the expected range of measurement method variables that might be encountered in normal use. The test is conducted using eight trials with factors set as indicated in the design table. The eight trial measurement results are used to calculate an estimate of the effect of each variable. (See Section 5.5, "Ruggedness Test Example.")

2.3 CALIBRATION AND UNCERTAINTY

Calibration is perceived by many users of measurement results to be the absolute establishment of accuracy or truth for the measurement system. *In reality, calibration is a statistical process that compares the measurement values from two or more measurement systems.* The first is the measurement system of a reference, or known laboratory. This is usually a "higher echelon" laboratory, often a standards laboratory that has been recognized in some formal way. The second is the measurement system being calibrated. In C&PI measurements, the comparison is often done using the measurement values obtained from both measurement systems for the "same material" referred to as "standard reference material" (SRM). Each portion of the SRM is considered to be the same. From sampling theory, there is uncertainty in any two portions of a sample being the same, at least with respect to the property being measured. From sampling theory, there is also uncertainty in any two portions of a *material* being the same. When samples are used for calibration, the requirement of homogeneity is critical. In reality, there is variability in the "standard reference material," and this contributes one component of the uncertainty in the standards laboratory's stated measurement result.

Because both the standard laboratory's measurement results and the measurement results of the measurement system being calibrated have uncertainties, the use of statistical regression is needed for proper calibration. The levels of uncertainty should be reported.[7] See Section 5.1, "Comparing Two Measurement Methods" and Section 5.2, "Calibration" for further discussion and examples of measurement system comparisons.

2.4 ROLE OF STANDARDS IN MEASUREMENT SYSTEMS

For the C&PI the word "standard" has more than one meaning. There are two main definitions we will consider here:

1. An official document that impacts or regulates how a laboratory operates by defining what is good practice.

2. A chemical or physical item (SRM) that is used to "standardize" or "calibrate" a measurement system as part of a control plan. In the C&PI there are many situations where there are no SRMs available that are traceable to a national standards body. We must then consider the development and control of "in-house" reference materials. (See Section 2.4.3, "Reference Materials.")

2.4.1 The Use of Standard Reference Methods

A standards reference method is an official document that impacts or regulates how a laboratory operates by defining the steps for making a specific measurement. These documents may be internal or external. In either case, the measurement methods must meet the criteria for a stable measurement system and provide fit-for-use measurement values. *Always evaluate the appropriateness of standard reference methods for what you are doing.* There are notable standard measurement methods on the books that give useless measurement values because the chemistry is not appropriate. This is most common with indirect (inferred) measurement methods, such as back titrations.

There are numerous sources for standard reference methods. They include international and national industry sector standards bodies, technical societies, and individual companies. The laboratory manager may need to know which markets (for example, aerospace, nuclear, automotive, or medical) the product is sold in to determine which standard measurement methods apply.

2.4.2 Measurement Laboratory Qualification

There are international standards such as ISO/IEC 17025 (formerly ISO Guide 25) and the ISO 9000 series. ISO/IEC 17025 is used to assess and accredit laboratories, and requires technical capability and the capability of a laboratory to produce measurement results that have well-characterized uncertainties and are traceable to National Measurement Institutes or other appropriate sources. These standards lead to a series of other standards that also affect laboratory operations:

VIM (International Vocabulary of Measurement), Second Edition, ISO, 1993

GUM (Guide to the Expression of Uncertainty in Measurement), 1995 (also available as ANSI/NCSL Z-540-2)

ISO/IEC 17025—First Edition 1999, General Requirements for the Competence of Testing and Calibration Laboratories

ISO Guide 38—General Requirements for the Acceptance of Testing Laboratories

ISO Guide 43—Development and Operation of Laboratory Proficiency Testing

ISO Guide 45—Guidelines for the Presentation of Test Results

ISO Guide 49—Development of Quality Manual for Test Laboratory

ISO Guide 55—Testing Laboratory Accreditation Systems—
General Recommendations for Operation

ISO Guide 58—Calibration and Testing Laboratory Accreditation Systems—General Requirements for Operation
and Recognition

ISO 10012-1—Quality Assurance Requirements for Measuring
Equipment—Management of Measurement Equipment

ISO 10012-2—Quality Assurance Requirements for Measuring Equipment—Guidelines for Control of Measurement Processes

ANSI/ASQC Q2-1995—Quality Management and Quality
System Elements for Laboratories Guidelines

This group of standards presents an overall view of good practices
related to the management of a laboratory. A C&PI laboratory should be
able to comply with all of the requirements, even if some of the terminology is not the same as generally used in C&PI. ISO 9000 series standards
are applied to laboratories when the laboratory is part of a production plant
or company registration process, or when an independent laboratory wishes
to demonstrate the competence of its quality system without demonstrating
technical competence to the same degree. Accreditation to ISO/IEC 17025
indicates the same level of quality system, as does registration to ISO 9002
in addition to technical competence.

National standards come from a variety of organizations. This discussion will not address them all. "Good Laboratory Practice" is addressed by
both the EPA and FDA in the U.S. Code of Federal Regulations (CFR):

40 CFR 160—Good Laboratory Practice Standards

21 CFR 58—Good Laboratory Practice For Nonclinical
Laboratory Studies

These standards apply to nonclinical studies and to measurement laboratories in that they require all data collected in the study to be collected in
laboratories that comply with the standards. This includes data such as chemical properties, physical properties, or data that determines the purity of the
material being measured. The laboratories involved must strictly adhere to
the study protocol once it is documented. In Europe, the laboratories providing
such data may have to be accredited. Currently in the United States, several
organizations accredit general-purpose laboratories to ISO/IEC Guide 25 or
17025. The most prominent ones are the National Voluntary Laboratory

Accreditation Program (NVLAP) and the American Association for Laboratory Accreditation (A2LA). Other U.S. accrediting bodies are not signatories to the international agreements that commit to mutual recognition of each others' work. Wide acceptance of results from labs that are accredited by U.S. bodies other than NVLAP and A2LA is not guaranteed.

There are many standards that are national and industry sector specific and impact laboratory operations. Industrial sector standards may address quality systems or they may be specific measurement methods required by government or industry specifications. Since these are specific to a market sector, they will not be discussed other than to say the laboratory must be kept current with the needs and requirements of the customer base.

ASTM standards are a compendium of best practices, measurement methods, and product specifications. The multivolume set addresses chemical measurements, instrumental methods, interlaboratory testing, and methods of data analysis.

The customer may expect (*or demand*) that the laboratory management has a thorough understanding of all requirements that may impact the validity of the measurement results produced.

2.4.3 Reference Materials

The hierarchy of reference materials is:

Acronym	Full Name	Primary Use
SRM	Standard reference material	Calibration
	↓	
SSM	In-house secondary standard material	Calibration
	↓	
RM	In-house reference material	Calibration
	↓	
CM	Control material	Measurement system control

An official standard reference material (SRM) is a primary standard traceable to a national body and is generally available commercially. The characterization of the material and the analysis methods are accepted nationally and perhaps internationally. These SRMs are accepted to have a "known" measurement value and statement of precision.

Since the SRMs tend to be expensive, they are often used to develop in-house working standards—secondary standard materials (SSM)—that are then used on a routine basis. An SSM is usually of the same chemical composition as the SRM.

It follows that if an SRM does not exist, a suitable in-house reference material (RM) should be prepared to serve the function of the SRM. An RM should be a stable homogeneous material. A characterization study should be run on an RM to determine the "known" measurement value and precision statement. See Section 3.5.3, "Controlling the Accuracy and Precision" for description of characterization studies. Often an RM is material taken from production during a time when it was in statistical control or a fully characterized quantity of conforming product.

A control material (CM) is a stable homogeneous material meant to represent, as closely as possible, the samples to be analyzed with regard to composition and physical properties. Usually, this is best achieved by taking a quantity of material from production during a time when it is in statistical control. The CM is used in daily work to collect data for monitoring and controlling calibrated instruments, measurement equipment, and the measurement method.

Maintenance samples are an instance of CM where portions of routine product samples are retained for later repeat measurements. Like all CM, maintenance samples must be homogeneous and stable when measured. Packaging, storage, and handling must not change the maintenance sample's properties.[9] See Section 3.5, "Controlling the Measurement System" for discussion of a measurement sampling program.

The users of these materials should be aware that in spite of confusion to the contrary, SRM, SSM, RM, and CM are completely separate materials with separate uses. Because the comparisons are between measurement values from the same measurement system, there is no requirement that a CM be "traceable." However, it is good practice to maintain traceability to the source from which reference materials were obtained. In contrast to CM samples, SRM, SSM, and RM are used to assure the accuracy of the analytical method as opposed to its stability. Measurement values of the analyte need to be well known, or at least highly precise, relative to the analytical method. The calibration also contributes to the final method's precision as explained by Hunter.[2] Frequency of calibration is driven by the stability of the measurement system as demonstrated by the CM control chart. When out-of-calibration occurs, good practice requires remedial action retroactive to the last record of in-calibration. All reference materials can be used to produce a control chart to statistically monitor the performance of the equipment and to detect statistically significant changes.

Recalibration may often be desirable after reestablishing control following an out-of-control situation. Such practices may reduce calibration frequency or give early warning of problems that might occur between scheduled calibrations. Interlaboratory comparisons (ILC) of measurement systems, intralaboratory control of measurement systems, maintenance sampling, and interlaboratory measurement control (IMC) are other common uses of reference materials. (See Section 3.5, "Controlling the Measurement System" and Section 3.5.5, "Controlling Measurement Systems on an Interlaboratory Basis.")

When a new analytical method is developed, a CM should be chosen for control charting the new method. The control chart provides information on the stability and precision of the analytical method. Out-of-control signals should initiate troubleshooting. Calibration should not be needed as long as the analytical method remains in control. The control chart should be the basis for continued confidence in the measurement results. In fact, calibration could cause a bias or out-of-control condition to appear in a measurement system if not done statistically correct.

An additional benefit from using SRM, SSM, RM, and CM is the removal of sampling variability allowing the measurement values to be used to estimate measurement variation. SSM, RM, and CM require that a procedure be developed and documented for preparation, purification, determination of purity, and use of these materials. Production records for each batch of the material must be documented. Sufficient data should be collected to allow the determination of the "known" measurement value, its precision, and the material's shelf life (stability). Shelf life should be determined for the material and stated as part of the reference material documentation. This increases our knowledge and control of these procedures, which leads to increased trust in the measurement results.

Good practice requires a written procedure for the initial introduction, periodic reevaluation, and replacement of any of the above reference materials. All reference materials should have a stated shelf life and be appropriately monitored from date of receipt.

A system for controlled distribution of the reference material (SRM, SSM, RM, and CM), complete with "known" measurement value and precision statement, needs to be developed. If the use of the reference material includes specifications where customers or suppliers would need the reference material, the controlled distribution would include customers and suppliers in addition to in-house distribution. The distribution procedures should include replacement materials on a periodic basis before shelf life expiration.[2, 3, 5]

Reference material characteristics

	SRM	SSM	RM	CM
Contains analyte of interest	√	√	√	√
Traceable	√	√		
Represents the composition of routine samples				√
Frequency of use	Yearly	Yearly	Yearly	Daily
Primary use	A	A	A	M
Known measurement value	√	√	√	
Statement of precision	√	√	√	
Homogeneous	√	√	√	√
Long-term stable	√	√	√	
Traceable to standards laboratory	√	√		
Source	1	1	1 or 2	2
Labeled	√	√	√	√
Expiration date	√	√	√	√
Not affected by packaging, storage, and handling	√	√	√	√

Yearly = Long term, greater than 30 days
Daily = Short term, 24 hours or less
A = Accuracy (calibration)
M = Measurement system control
Labeled—see Section 2.2.3.10, "Labeling Samples."
Source—1 = commercially available; 2 = an existing product lot

2.5 ENVIRONMENTAL EFFECTS ON THE MEASUREMENT RESULT

Two environments are key:

1. The environment in which the sample is taken; and

2. The environment in which the measurement value is obtained.

All environmental factors can have an influence—temperature, light, air, barometric pressure, humidity, and so on. Any impact must be quantified and taken into account when taking the sample, obtaining the measurement value, and calculating the measurement result.

The environment for sampling must not affect the sample or the bulk material. For example, a material that absorbs water should be sampled under conditions that neither allow removal of water from either the material or sample nor allows the absorption of more water by either the material or the sample. The environment of the measurement method frequently causes variation in the measurement result. For example, boiling point measurement values vary with barometric pressure, so they are always corrected for barometric pressure.

It is not always possible to eliminate the impact of environmental factors. Furthermore, environmental factors are not always cost effective to control. This requires their impact to be quantified and taken into account throughout the measurement system, especially when calculating both measurement results and measurement uncertainty.

2.6 MEASUREMENT SYSTEM MATERIALS

Many measurement systems employ materials to prepare, preserve, transport, or dissolve the sample. While these materials are not intended to be measured, they can have a profound effect on the measurement values. Impurities can raise or lower measurement values. One of the common ways of quantifying the impact of measurement materials is the use of material blanks. A material blank without the sample is run through the measurement system to quantify the measurement values due only to the measurement material. Blanks do not quantify the impacts of components in the blank that react with the analyte in the sample. The properties of measurement materials are critical and must be quantified, including the material's variability.

Other consumable materials can also have an effect on the value of measurement results, even though they do not participate in any chemical processes. For example, when making up an ice bath as a temperature reference for zero degrees Celsius, impurities in the water used will affect the freezing point and, therefore, the accuracy of the reference temperature.

ISO/IEC 17025 requires control over both types of materials, in fact, over any consumable material that can have an effect on the quality of measurements. Don't, however, confuse consumables with reference materials, which have separate and different requirements. A reference material is one whose actual chemical or physical property is the parameter of interest (such as a standard viscosity or conductivity solution). The properties of a consumable will affect measurements, but will do so indirectly as in the previous ice example.

2.7 MEASUREMENT SYSTEM VARIABLES

Variables of measurement systems include:

- The environmental conditions at the time the sample was taken. (See Section 2.5, "Environmental Effects on the Measurement Result.")

- The storage conditions of the sample. (See Section 2.2.3.9, "Sample Retention—Packaging and Storage.")

- The sample preparation conditions.

- The settings of equipment, sequence of steps, duration of steps, and environmental conditions when the measurement method was run. (See Section 2.5, "Environmental Effects on the Measurement Result.")

Measurement system variables must be appropriate and show that the measurement system was in control. (See Section 3.5, "Controlling the Measurement System.") As with variation in environmental factors discussed in Section 2.5, variation in these other conditions impact the measurement value and the measurement result.

Variables are those conditions that we have control over and can set to a specified value. For example, the injection port temperature of a gas chromatograph is a variable that can be set to the temperature specified in the measurement method. While the standard gas chromatograph port temperature of a lab is 200° C and very stable, it is not appropriate when the analyte to be measured polymerizes at 150° C. Neither is a 20 minute run time appropriate for a gas chromatograph when an impurity comes off the column at 30 minutes, which is 15 minutes after the main peak.

As with production process variables, there is variation in measurement system variables and they must be quantified and controlled if measurement values are to be trusted. (See Section 2.2.8, "The Need for Ruggedness Tests.")

2.8 DOCUMENTING THE MEASUREMENT SYSTEM

Documenting the measurement system is an essential step in delivering trustworthy measurement results. Volumes have been written on the importance of documenting to reduce variability. A trustworthy measurement result is seldom delivered by an undocumented measurement system. See Section 2.4.2, "Measurement Laboratory Qualification" for standards that can be used as guides to documentation.

2.9 CHECKLIST FOR TRUSTWORTHINESS

Item	Book section	Rating
Decision criteria defined.	2.2.1	
Quality system documented and implemented—certifications?	1.8	
Measurement method can detect the difference needed to make the decision.	2.2.2	
Chemical reaction is specific.	1.7	
Physical reaction is specific.	1.7	
Unit of measure is suitable for decision.	2.2.1	
Measurement result is reported with statistical information.	2.2.7	
Sampling procedure documented.	2.2.3	
Person taking sample is trained.	2.2.5	
Sample is representative.	2.2.3.4	
Bulk material was homogeneous.	2.2.3.5	
Random sample taken.	2.2.3.8	
Samples containers nonreactive.	2.2.3.9	
Sample storage did not change material.	2.2.3.9	
Sample labeled correctly.	2.2.3.10	
Sample chain of custody intact.	2.2.3.11	
Accredited measurement laboratory.	2.4.2	
Measurement system documented.	2.8	
Method of controlling measurement system.	3.5	
Control chart on measurement system.	3.5.2	
Precision and accuracy.	3.3.1	
R&R study done.	3.3.2.2	
Ruggedness test done on measurement method.	2.2.8	
Equipment calibration status.	2.3	
Type of reference standards used in calibration.	2.4.3	
Standard measurement method(s) used.	2.4.1	
ILC used.	3.5.5	
Blind samples used.	3.5.4	
Outlier procedure documented.	3.6	

(continued)

Item	Book section	Rating
Environmental conditions at time of measurement.	2.5	
Measurement system material effects.	2.6	
Measurement system variables.	2.7	

Ratings: T = Trust, NT = No trust
Any NT should trigger corrective action.

This checklist is provided as a quick reference to help you answer the question: After all is said and done, should you report or accept the measurement result? This checklist encompasses the viewpoints of the user of the measurement result, generator of the result, and management of both areas.

We now have a trustworthy measurement result!

2.10 REFERENCES

1. W. J. Youden, *Statistical Techniques for Collaborative Tests* (Gaithersburg, MD: International Association of Official Analytical Chemists, 1967).
2. J. Stuart Hunter, "Calibration and the Straight Line: Current Statistical Practices," *Journal of the Association of Official Analytical Chemists* 64, no. 3 (1981): 574–583.
3. George A. Uriano and J. Paul Cali, "Role of Reference Materials and Reference Methods in the Measurement Process," in *Validation of the Measurement Process*, ed. James R. DeVoe (ACS Symposium Series No. 63, Washington DC, 1977).
4. AOAC International, *Intralaboratory Analytical Method Validation* (Gaithersburg, MD: International Association of Official Analytical Chemists, 1977): 4-1–4-4.
5. ASTM, E826-85, *Standard Practice for Testing Homogeneity of Materials for the Development of Reference Materials* (West Conshohocken, PA: American Society for Testing of Materials, 1996).
6. ASTM, E1169-89, *Standard Guide for Conducting Ruggedness Tests* (West Conshohocken, PA: American Society for Testing of Materials, 1996).
7. International Organization for Standardization, *Guide to the Expression of Uncertainty in Measurement* (Geneva, Switzerland: International Organization for Standardization, 1993).
8. Susan E. Daniels, assoc. ed., "Tire Failures, SUV Rollovers Put Quality on Trial," *Quality Progress* 33, no. 12, (December 2000): 30–46.
9. J. M. Juran, *Quality Handbook*, Fifth edition (New York: McGraw-Hill, 1998).

3

Using the Measurement Value and Measurement Result

A partial list of valid uses of measurement values and measurement results:

- Measurement system control
- Production process control
- Product disposition—conformance to specifications
- Troubleshooting
- Interlaboratory comparisons (ILC) of measurement systems
- Government regulatory compliance reporting
- Customer complaint resolution
- Understanding variation

Of these, the following sections discuss:

- Measurement system control
- Troubleshooting
- ILC
- Understanding variation
- Other uses of measurement results

3.1 THE MEASUREMENT RESOLUTION NEEDED TO MAKE THE DECISION

In each of the above uses, the measurement system precision needs to be better than the precision of the process under consideration. A discrimination ratio[29] of four is a minimum for making a correct decision. The formula for discrimination ratio is:

$$Discrimination\ Ratio = \sqrt{\frac{2 \times \sigma^2_{process} + \sigma^2_{test)}}{\sigma^2_{test}}} - 1$$

A measurement system has resolution if its variability is small enough to differentiate the smallest difference needed to make the decision. Knowing the decision criteria that will use the measurement result is essential to determining if a measurement system has the required resolution. (See Section 2.2.1, "The Decision and Unit of Measure Must Be Defined.")

For example, to decide if a material meets specification, the measurement resolution needs to be enough to make a correct accept/reject decision with the required confidence. If the decision limit reads 50 ± 10 mm, the range is from 40 to 60 mm. Using ASTM E29:93A, Section 6.4, Rounding Rules, 39.5 mm would be rounded to 40 mm and 60.5 mm is rounded to 60 mm. The measurement resolution must ensure that we can differentiate between 39.5 mm and 39.4 mm, and between 60.5 mm and 60.6 mm, because 39.4 and 60.6 are outside of the limits. Thus the unit of measure for the decision should be 0.1 mm. The measurement must be capable of detecting differences of 0.1 mm.

The uncertainty associated with the measurement result needs to be established. Which type of number gives the differentiation needed for the decision? Is it an average, standard deviation, ratio, and so on? How many measurement values should be used in these calculations? Determining the necessary differentiation requires production process and product knowledge. The uncertainty is as much a part of the measurement result as the unit of measure is a part of the measurement result. Examples: 33.5 g (average of 4) ± 0.3 (one standard deviation of individual averages) is a correctly stated measurement result; 15.5 g ± 0.2 (two standard deviations of individual measurement values) is also correct; 33.5 g and 15.5 g by themselves are incomplete and could be misinterpreted by the user of the measurement result.

3.2 COMPARING MEASUREMENT RESULTS TO DECISION LIMITS

Quantifying of uncertainty in equations and calculations is a misnomer. Uncertainty resulting from the original measurement may be carried or propagated throughout calculations. The amount of carryover depends on the effect of independence and covariance in the equation.

ASTM E29, *Using Significant Digits in Test Data to Determine Conformance with Specifications,* describes two common methods used to compare measurement results with specifications, tolerance limits, or acceptance limits. The *absolute method* implies that any deviation, however small, outside the limit signifies nonconformance. If the limit was 2.5 ppm maximum, then 2.501 ppm would be out-of-limit. Absolute limits of 2.5, 2.50, and 2.500 are all treated the same. The *rounding method* implies that a measurement result shall be rounded to the same number of decimal places as the limit, and then compared to that limit. If the limit was 2.50 ppm maximum, then 2.505 ppm would be rounded to 2.50 and be in-limit. However, if the limit was 2.500 ppm maximum, then 2.505 ppm would be out-of-limit. Rounding limits of 2.5, 2.50, and 2.500 are all treated differently.

It is therefore appropriate that a listing of limits include a qualifying statement such as "Limits are absolute limits" or "Measurement results shall be rounded to match the number of decimal places of the limit before comparisons are made." This entire field is in the throes of change, in large part because of our increased understanding of the uncertainty in measurement results. See Section 4.3 for a discussion of what the 21st century may look like in this field.

3.3 CHARACTERISTICS OF A TRUSTWORTHY MEASUREMENT SYSTEM

The measurement values of a good measurement system show that it is Stable, Precise, Accurate, Representative, Site comparable, and Effective (SPARSE).

Stable

A measurement system can be shown to be stable (does not drift) using a measurement system control chart on which measurements of a control material (CM) are plotted. The measurement system control chart could consist of individual measurements and a moving range, or involve multiple measurements of the CM. See Section 3.5, "Controlling the Measurement System," for further discussion of control charts for measurement systems.

Precise

A measurement system can be shown to be precise enough for its intended use by using a control chart, where product measurement results are plotted and the control limits are based on the estimate of the measurement system variability calculated from the range of multiple measurements of each sample. If the measurement system is precise enough (discrimination ratio 4 or greater), the ranges will be small, the control limits will be tight, and the $\overline{X}$ chart should appear to be out-of-control. (See Sections 3.3.2.2, "Repeatability & Reproducibility (R&R) Studies" and 3.3.2.3, "Repeatability and Reproducibility from Control Charts.")

Accurate

The accuracy of a measurement is established through calibration. Continued accuracy can be demonstrated using traceable (SRM or SSM) or internal reference materials (RM or CM) for calibration checks or standardization. (See Section 2.3, "Calibration and Uncertainty.")

Representative

A sample can be shown to be representative of the original population (drum, truckload, railroad car, and so on) by comparing the measurement results of a homogeneity study with the existing sampling procedure. (See ASTM E826, *Testing Homogeneity of Materials for the Development of Reference Materials* and Section 2.2.3.2 through Section 2.2.3.5 on sampling.)

Site Comparable

A measurement system can be shown to be site comparable when the measurements agree within a specified level of confidence among laboratories located at different independent sites. (See Section 3.5.5, "Controlling Measurement Systems on an Interlaboratory Basis.")

Effective

A measurement system can be shown to be effective if it is adequate for its intended use. The most cost effective number of measurement results needed to make an informed decision should be based on the cost of measuring and the cost of making a wrong decision concerning the disposition of the material. (See Section 2.2.2, "The Measurement Method Must Be Suitable for Its Intended Use.")

3.3.1 Measurement System Precision and Accuracy

The ability to ensure adequate precision and accuracy (lack of bias) of a measurement system should be the primary concern of laboratory quality assurance. The first step is to require that the measurement method development process includes determining both precision and accuracy. It may seem trivial to state that precision and accuracy cannot be controlled or improved unless they have been determined, but too often laboratory personnel do not engage in the effort to establish precision and accuracy of a measurement method until after the reliability of the measurement results have been questioned. There are no shortcuts to the work that must be performed to establish both the precision and the accuracy of a measurement method. Resources and time must be made available to complete the work necessary to satisfactorily determine both the precision and accuracy of a measurement method. Methodologies for development of estimates of precision and accuracy of measurement methods are described in ISO 5725; ASTM E180, *Determining the Precision of ASTM Methods for Analysis and Testing of Industrial Chemicals*; and the United States Pharmacopoeia among others.

3.3.2 Measurement System Variation

The variation observed in the measurement results of finished material contains all of its history, including variability from the production process itself as well as the variability from the measurement system. Analysis of variance techniques can separate this variability into its components. In any case, it is important to know the portion of the total variability that comes from the measurement system. The amount of measurement variability that is acceptable depends on how the measurement results will be used, and would be different for production or measurement system control, material characterization, or material identification.

Determining the measurement variability does not require NIST (National Institute for Standards and Technology) traceable standards, such as those used for equipment calibration. The variability is best determined by using CM that is typical of the material being measured. (See Section 2.4.3, "Reference Materials.") This requires a quantity of uniform material large enough to do the number of measurements needed to estimate the variance. It is important to assess the variability of the overall measurement system.

After obtaining a good estimate of the measurement variability, the estimate can be used to determine sample size, control limits, and so on for a control chart of the measurement system. Variability estimates are usually valid over the typical range of measurement results expected for the material. This means that the sample size, control limits, and so on to be used in the future can be based on previous data.

Since it is common to use measurement results as a means to classify product or evaluate production process performance, it is essential that the measurement system produce the "right" answer. In any type of measurement, it would be desirable to have the variability of the measurement system be:

- Small compared to any other nonmeasurement system variation, for example, the actual variations of the production process; and

- Able to depict common cause variation, that is, in the state of statistical control.

In addition, the measurement instrument capability (expressed as $6\sigma_{instrument}$, see Section 3.3.2.2, "Repeatability & Reproducibility [R&R] Studies") should be small relative to the specification range. Measurement capability divided by the specification range is often referred to as the precision-to-tolerance ratio (P/T) and should be smaller than 0.1.[1] To appreciate the havoc that measurement variability can reap, you only need look at the U.S. Presidential election of 2000. The amount of measurement variability in the balloting processes of several states made it scientifically impossible to declare a winner in those states. The ensuing legal actions are a testament to U.S. democracy and rule of law. The actions are also a testament as to what can happen when people do not understand what measurement variability is, how it affects decision making, and what to do when it cannot differentiate between options (candidates).[40]

In order to overcome the concerns and issues pointed out in sections 3.3.2.2 and 3.3.2.3 with respect to R&R studies in the C&PI, it is necessary to use techniques to continually estimate the measurement system variability components. Section 5.7, "Continually Updating Estimates of Measurement System Variance Components," gives an example for such a technique.

3.3.2.1 Monitoring and Control of Measurement Precision

The precision of a measurement method has two components: repeatability and reproducibility. Repeatability is considered short-term, or within sample variability, where everything is kept as constant as possible: same instrument, same operator, same time of day, same room temperature/ humidity, and so on over a short period of time. It is a measure of the minimum possible variability associated with a measurement system.

Reproducibility is variability where everything is varied as much as possible while still following the correct procedure under typical operating conditions: different instruments, different operators, different times (work shifts), different seasons (temperature/humidity), and so on over a long period of time. It is an indicator of the maximum typical variability of a measurement system.

3.3.2.2 Repeatability & Reproducibility (R&R) Studies

The procedures used in the typical Gage R&R studies were primarily the outcome of efforts in the automotive industry to evaluate the capability of gages and operators in a production environment. For the C&PI, we will refer to these studies as R&R studies. The fundamental issues that this area addresses are whether the measurement system:

- Is stable;

- Has adequate discrimination; and

- Has relatively small variation.

Assessing measurement system variation often focuses on evaluation of instrument capability. Control charts are helpful for separating (and estimating) the total variation observed into those components relating to the material measured and those relating to the instrument used:

$$\sigma^2_{total} = \sigma^2_{material} + \sigma^2_{instrument}$$

Once we can separate the total variation into material and instrument components, we can begin to further subdivide the instrument variation, $\sigma^2_{instrument}$, into its two components of repeatability and reproducibility. Repeatability reflects the inherent variation of the gage itself, while reproducibility is the variation due to different operators or technicians using the instrument. The former is the variation *within* a single measurement and the latter refers to the variation due to differences *between* operators. Another important aspect of these two sources is time. Generally, these components deal with what is referred to as short-term variation.[3] (See Section 5.7, "Continually Updating Estimates of Measurement System Variance Components.")

One of the major differences between C&PI and industries such as automotive and parts manufacturers is the element of time and its ramifications on sampling. In the C&PI, production processes often drift over time and data are often autocorrelated. Data from these production processes are usually based on individual measurement values (subgroups of size one) and many production processes are not in the state of statistical control. Short- and long-term components of variation from the production process and measurement systems are important. Another concern is specifications used in the C&PI that are often not statistically based on production process and measurement system knowledge.[3] Without this knowledge, comparisons of production process capability with specification ranges are questionable.

3.3.2.3 Repeatability and Reproducibility from Control Charts

In the automotive industry, procedures for evaluating repeatability and reproducibility are described in the A.I.A.G. reference manuals for measurement systems analysis.[2] Similar procedures can be used for the C&PI. The basic procedure consists of randomly taking several units of production (units of product) and having the production process operators or lab technicians who would normally take the measurements use the instrument to measure each product unit several times. The data is then put into control chart format ($\overline{X}$ and R charts), and these charts are used to calculate the standard deviations due to repeatability and reproducibility.

As an example, consider the following data from Montgomery:[1]

Part number	Measurement 1	Measurement 2
1	21	20
2	24	23
3	20	21
4	27	27
5	19	18
6	23	21
7	22	21
8	19	17
9	25	23
10	25	23
11	21	20
12	18	19
13	23	25
14	24	24
15	29	30
16	26	26
17	20	20
18	19	21
19	25	26
20	19	19

For this C&PI example, let's assume that the part number refers to the actual sample number for a viscosity measurement on some chemical, and the operator or lab technician is taking enough sample to run two measurements on each. In this case, one would like to have the operator who is taking the samples be the same person doing the viscosity measurements. This limits the sampling of the 20 pieces to one shift (8–12 hours) of the production process, unless the evaluation period is extended into overtime. Realistically, the sampling period would cover the entire measurement period.

The R&R study and analysis follow the same procedure shown in Montgomery.[1] As Montgomery points out, the resulting $\overline{X}$ and R charts for the 20 samples are very informative. The $\overline{X}$ chart could have out-of-control points (unless the measurement system is very consistent). Out-of-control points are an indication of the ability of the instrument to discriminate between samples. He goes on to say that the R chart shows the magnitude of the measurement error. In this example, the R chart is in-control, indicating that the operator has little difficulty in making consistent measurements. If it were out-of-control, Montgomery[1] points out that this would indicate the operator was having difficulty using the instrument. The $\sigma^2_{instrument}$ is estimated from $\overline{R}$ (R chart) and σ^2_{total} can be estimated by calculating the standard deviation of the 40 readings of viscosity and squaring it. Then as shown in Montgomery,[1] the variance and standard deviation of the product viscosity can be estimated by calculation using the equation given in section 3.3.2.2, "Repeatability & Reproducibility (R&R) Studies."

The reference example[2] is expanded to show how estimates of the repeatability and reproducibility components of variance for this measurement system can be obtained. The extension involves repeating the same study described above with multiple operators or lab technicians using the same instrument. Once again, the sampling is limited to a logical time period of production and subsequent measurement. The reference[2] gives the data for three operators and 20 samples and shows the simple calculations for estimating repeatability and reproducibility. As expected, the standard deviation of measurement error increases (multiple operators). If the largest source of variation were reproducibility, then one would train the operators to produce more uniform measurements. If repeatability is the larger component, one should consider changing the instrument (measurement system).

In the C&PI, the estimation of variance components for product and measurement may not be easy. On the product side, lot definition may not be clear. This is particularly so if the production process is continuous or semicontinuous. Should a lot be defined as the product produced in an eight hour shift or do we need to define the lot based on smaller time increments (four hours) because the within-lot variation is large? Then there is the question of how many lots to sample. One suggestion is a minimum of 30, with 60 preferred.[3] Of course, the sampling issues pointed out in earlier sections apply in these cases. The duration of the evaluation or the actual production time sampled is also a concern. Once again, a minimum of 90 days is recommended.[3]

The measurement system components have similar issues. Should the technicians measure each sample on the same day or different days? Can one ensure that repeated measurements are performed under the same conditions? Again, sampling issues are important.

These concerns show some of the major differences between Gage R&R studies in the piece- and automotive-type industries, and R&R studies in the C&PI. In general, it can be said that in the C&PI many more samples should be taken over a much longer time period to insure improved and more meaningful estimates of the sources of variation for product, repeatability, and reproducibility. The underlying problem in using R&R studies, as discussed, has it roots based on the small degrees of freedom associated with the estimated variance components and the short duration of the studies. As stated earlier, this can only be improved by increasing the number of samples and the duration of the study. For appropriate procedures to address these improvements, see Section 5.7, "Continually Updating Estimates of Measurement System Variance Components."

3.4 WHY COMPARE TWO MEASUREMENT METHODS?

There is more need to compare two measurement methods than might be thought at first glance. Even when the same documentation is being followed, personal interpretation based on training can result in widely different measurement methods. Consider the number of interpretations of a method document that contains terms such as hot, cold, fast, make acid, slow, and quickly by several trained laboratory technicians. There could be as many measurement methods due to such interpretations as the product of multiplying the number of people, times the number of ambiguous terms, times the number of possible interpretations of each term.

There is a need to compare the precision of comparable measurement methods in the same laboratory. Examples:

- A new and less costly measurement method has been developed that is expected to give equivalent measurement results. The proposed measurement method needs to be compared with the currently used measurement method for both level and precision.

- A new measurement method has been developed that is claimed to provide better precision than the current measurement method. One method uses infrared and the other uses ultraviolet to detect the analyte. A comparison between the methods is needed to quantify the correlation between the two methods of detection in order to make the necessary changes to specifications, acceptance procedures, control charts, and so on.

- A proprietary measurement method has been developed that is expected to give equivalent measurement results to an ASTM method. A comparison is needed between this method and the appropriate ASTM method. The calibration methods for the two measurement methods also should be compared.

More specifically, there is always a need between supplier and customer to develop the correlation between two different measurement methods being used in different laboratories to report the measurement result of the same property in the same material. Examples:

- Your supplier uses Karl Fisher for water and you use infrared. You need to set a specification for using infrared, so you need to compare the measurement methods.

- You are getting different measurement results from different laboratories on different samples from the same bulk material. You need to compare the two laboratories, operators, measurement equipment, and so on. You are interested in both measurement precision and property level.

- In order to accept product based on online measurement results, you need to compare the laboratory method with the online method.

- A referee laboratory is auditing one of your environmental quality measurements. You would like to compare your measurement method with the referee method.

To compare two measurement methods, a sampling experiment is designed with samples covering the expected range of measurement values and measurement results that will be encountered. Repeated measurements are made on each sample by each method. In all cases, it is important to have a wide range of real differences among the samples used to generate the data.

It is important to note that standard least squares regression is not appropriate for comparing two measurement methods. Standard regression techniques assume that the X variable is known and is without error. When comparing two measurement methods, both X and Y are subject to error. See Section 5.1, "Comparing Two Measurement Methods" for an explanation and an appropriate procedure. An important application of measurement system comparison is described in section 3.5.5, "Controlling Measurement Systems on an Interlaboratory Basis."

3.5 CONTROLLING THE MEASUREMENT SYSTEM

Controlling the measurement system requires discipline. A good measurement system includes a procedure for monitoring the everyday performance as well as periodic audits. Documented, sound procedures and adequate training are required to control the measurement system. The measurement system should be a documented component of a functioning quality system. (See Section 1.8, "Quality System's Role.") One of the most appropriate means to monitor and control a measurement system is the use of measurement system control charts. This practice requires continual monitoring of the measurement system. It is best achieved by measuring a control material (CM) repeatedly at defined intervals and charting the measurement results on a control chart. It is good practice to submit at least some of the CM as blind samples. The control chart can be of several types, including Shewhart $\overline{X}/R$, EWMA, CUSUM, zone chart, and so on.

Four common programs that produce data useful in controlling and improving measurement systems are:

1. Intralaboratory control of measurement systems

2. Interlaboratory comparisons (ILC) of measurement systems

3. Interlaboratory measurement control (IMC)

4. Maintenance sampling

This matrix shows what is common among these programs.

Program	Measurement system sites	Reference material used	Performance interval	Use of data
Intralaboratory	1	CM	Short term— daily, shift, less than 24 hours	Measurement system control
IMC	2 or more	RM	Short term— daily, shift, less than 24 hours	Measurement system control
ILC	2 or more	RM	Long term— annually, semi-annually, biennially	Auditing, variance component estimation
Maintenance sampling	1	CM	Short term— daily, shift, less than 24 hours	Estimating VLL, VPP, VWP, VLT, VST

See Section 2.4.3, "Reference Materials" for discussion of RM and CM. See Section 5.7, "Continually Updating Estimates of Measurement System Variance Components" for definitions of VLL, VPP, VWP, VLT, and VST.

Intralaboratory control of measurement system refers to the statistical process control being used within one site to make the decisions on whether or not a measurement system is off aim at any given moment. The CM used is local to the particular site where such control is in place.

An IMC is the external version of intralaboratory measurement control. Samples from one supply of RM are used at each measurement system site involved. When a CM is used in an IMC it must be elevated to the characteristics of an RM by appropriate analysis and procedures. This facilitates management of the measurement system on a broader basis, including quick detection of biases and timely initiation of improvement activities. If site averages are plotted over time, with all sites appearing on the same plot, it is straightforward to include various statistics with the plots, such as current biases and current estimates of measurement standard deviations, in order to keep the measurement system information up to date. The data can be analyzed for long-term variation, such as is done in ILC.

ILC are the long-term variation analysis version of IMC. The RM supply for ILC is normally different from the RM supply for an IMC, primarily to facilitate management of the two programs. It is much easier to keep them separate when the RM supplies are different than when only one RM supply is used. ILC can serve as audits of an existing IMC or as the primary sources of information on long-term variation and biases among sites, if no IMC is in place. See Section 3.5.5, "Controlling Measurement Systems on an Interlaboratory Basis" for more detail on ILC. Also, see Section 5.6, "A New Measure for Measurement System Improvement Efforts," which describes the use of ILC to provide rich information for diagnosing measurement system problems and for hints at where to look for a fix.

Maintenance sampling is one way of estimating measurement system variation by measuring a CM repeatedly over time, then examining and evaluating measurement results and distribution by appropriate statistical treatments—control charting, and so on.[5] For maintenance sampling, the CM is a retained routine product sample. (See Section 2.4.3, "Reference Materials" and Chapter 6, "Glossary.") The data are collected daily and analyzed periodically, such as quarterly, to estimate all sources of variability in a product property (characteristic), including both short- and long-term product variability and short- and long-term measurement variability. The resulting variance components can be used to help decide which product and/or measurement sources of variability should be reduced. (See Section 5.6, "A New Measure for Measurement System Improvement Efforts.")

The business purpose of a measurement system is to deliver measurement results for business decisions, such as production process control and product release. To do this, the measurement system must use measurement results from SRM, SSM, RM, and CM to establish and maintain control of the measurement system. A measurement system's existence depends on the support it provides to the business.

3.5.1 Relative Sensitivity of Measurement and Production Control Charts

The measurement system control chart needs to be designed to detect a shift in the measurement system average before the production process control chart can detect it. Otherwise, there is a potential for a production process adjustment to be made because the measurement system is off aim even when the production process is really on aim.

You can use the average run length (ARL) curve to evaluate the relative sensitivity of the measurement and production process control chart designs. Example:

Production process control chart $s = 3.2$
Sampling frequency = once per shift

Measurement system control chart $s = 2.1$
Sampling frequency = once per day

Both control charts using Shewhart with American Telephone and Telegraph (AT&T) runs rules.[11] From the ARL curve for the production process, a shift of one standard deviation is a shift in the production process average of 3.2 units. The average run length is 9.22 time units, or 73.76 hours. A shift of 3.2 units on the measurement system control chart is $3.2/2.1 = 1.52$ measurement standard deviations.

From the average run length curve:

Production process control chart (Shewhart)
Sample frequency = 1 per 8 hour work shift
Time to detect one standard deviation shift = 9.22 samples, or
      74 hours

Measurement system control chart (Shewhart)
Sample frequency = 1 per 24 hour day

Find the ARL for a shift of 1.52 standard deviations: Using linear interpolation, the ARL for a shift of 1.52 standard deviations in the measurement system control chart is 4.81 samples, or 115 hours.

Average run length curve for Shewhart with AT&T runs rules:

Shift	Samples	Hours
1.2	6.89	
1.4	5.41	
1.52	**4.81**	**115**
1.6	4.41	
1.8	3.68	

Clearly in this case, the measurement system control chart is not sensitive enough to detect shifts earlier than the production process control chart. If a shift of 3.2 units occurs in the measurement system, the production process control chart would detect it in 74 hours, but the measurement system control chart would not detect it for 115 hours.

Assume that the goal is to detect the one standard deviation shift in about half the time. There are two ways to achieve the goal:

1. Increase the sampling frequency of the measurement control materials.

   For this example, increasing the sampling frequency to once per 8 hour work shift will result in an average run length of about 38 hours (4.81×8h).

2. Increase the daily sample size to reduce the measurement control standard deviation.

   For this example, increasing the sample size by a factor of 4 will reduce the standard deviation to 1.05:

$$\text{new } s = \frac{\text{current } s}{\sqrt{4}} = \frac{2.1}{2} = 1.05$$

   The ratio of the production process standard deviation to the measurement system standard deviation is now 3.2/1.05 = approximately 3.0.

   From the ARL table, the average run length for a shift of 3.0 is 1.67.

   For a once daily set of four per 24-hour day sampling frequency, this relates to an ARL of about 40 hours.

Either of these options is satisfactory. The increase in sampling frequency to once per shift causes a three-fold increase in sampling. The increase in sample size causes a four-fold increase in sampling. It now becomes a nonstatistical problem to decide which option to adopt.

An additional control chart can be used to monitor the sensitivity of the measurement system control charts. One suggestion involves plotting the data used for the production process control chart on a control chart equivalent to the measurement system control chart. The production process data should generate off-aim signals on the measurement system control chart even though the production process control chart is in-control.[10–12, 25] (See Section 3.5.6, "Choosing What to Improve.")

3.5.2 Measurement System Control Charts

Two primary objectives of measurement system control charts are:

1. To ensure that the measurement system is continuing to produce accurate measurement results with the required precision.

2. To ensure that changes in the measurement system are detected with the measurement system control charts before the equivalent changes are detected with the production process control charts. For example, if there is a shift in the measurement values due to a loss of calibration, this shift will also be present in the measurement results of the routine product samples. If this shift is large, it may cause an out-of-control point on the production process control chart when the production process is actually in-control.

The basic Shewhart $\overline{X}$ and R control chart is probably the most commonly used control chart and has the most literature supporting its use. There are several modifications in the criteria for detecting an off-aim situation that are useful. For example, an out-of-control or off-aim signal is given by the following run rules for these Shewhart type modifications:

Shewhart	1 point outside of ± 3 *s*
Shewhart (Ford)	1 point outside of ± 3 *s* 7 consecutive points on the same side of aim
Shewhart (Ott)	1 point outside of ± 3 *s* 2 consecutive points outside of ± 2 *s* 7 consecutive points on same side of aim
Shewhart (AT&T)	1 point outside of ± 3 *s* 2 of 3 consecutive points outside of ± 2 *s* 4 of 5 consecutive points outside of ± 1 *s* 8 consecutive points on same side of aim

These four options each have different average run length curves that estimate how long it takes, on average, to detect shifts in the measurement system average expressed as multiples of standard deviation. The ARL curve for the Shewhart (AT&T) chart closely matches that for the CUSUM chart, except for very small shifts in the average where the CUSUM results in fewer false out-of-control signals. (See Section 3.5.3, "Controlling the Accuracy and Precision.") Therefore, either is a good alternative for detecting shifts from aim. The choice for a specific operation may depend on the familiarity with using the type of chart and/or computer software availability at that location. Both Shewhart and CUSUM control charts are included in most good SPC software. The CUSUM chart has additional advantages that can enhance the control of the measurement system:[14-18]

- The CUSUM procedure provides an estimate of a process shift when an off-aim signal occurs.

- A CUSUM procedure provides an estimate of the time when the shift occurred.

- Matching the measurement system control performance to that of the production process control can be done more precisely with the CUSUM procedure by changing either or both of the two design parameters (h and k).

- The CUSUM plot is a very useful tool for identifying times when changes occur. If a CUSUM plot were made for other variables, one would look for changes in one chart that correspond to changes in other charts to aid in identifying possible assignable causes.

A brief example on how to design a basic CUSUM control system follows. Three important pieces of information are required for CUSUM:

1. Aim—the target value for the control chart. This is usually the predetermined average for the measurement control sample supply.

2. s—the standard deviation of the measurement values to be entered into the control chart. This is usually a predetermined estimate from either the subgroup ranges or the average moving range.

3. Δ—the difference from aim that you would like to detect in a reasonable time. For the basic CUSUM. This is usually set equal to one standard deviation.

From this information we design the basic control chart, where $h = 4.0$, $k = 0.5$.

Term	Value
Center line	Aim
$H = hs$	4.0 s
$K = k\Delta$	$\Delta/2$
Action limits ($\pm$ H)	$\pm$ 4.0 s
Slack ($\pm$ K)	$\pm \Delta/2$

For each data point X_i, compute:

SH = Cumulative sum of differences above aim:

$$SH = \text{Max} \ [0, (X_i - \text{Aim}) - K + SH_{(i-1)}]$$

NH = Number of successive points with a value for $SH > 0$.

SL = Cumulative sum of difference below aim:

$$SL = \text{Max} \ [0, -(X_i - \text{Aim}) - K + SL_{(i-1)}]$$

NL = Number of successive points with a value of $SL > 0$.

An out-of-control signal is given when either SH or SL exceeds the value of H.

(See Section 5.3, "Example of CUSUM" and references 15–18.)

3.5.3 Controlling the Accuracy and Precision

To control the accuracy and precision of the measurement system, one begins by preparing a large stock of measurement reference material (RM). The material should be from an actual product lot, and should be made as uniform as possible without changing the properties of the material. Nonliquid materials, such as granules or fibers, should be well mixed or blended to provide uniformity in the measurement control stock. Obtain enough measurement values over a period of time to provide an initial estimate of the standard deviation and aim using the subgroup range or average moving range from individual data. This characterization of the stock is referred to as a characterization study and it is recommended that at least 30 sampling intervals (individual data) be used.

Two useful control chart methods for measurement system control are the Shewhart control chart using the American Telephone and Telegraph (AT&T) runs rules and the CUSUM control chart. The average run length to detect a shift in the measurement system average is quite similar for shifts larger than 0.5 standard deviations. The following shows the average run-length curves.

Average Run Lengths (ARL) for Shewhart and CUSUM control charts estimate how long it takes, on the average, to detect shifts in the measurement system average expressed as multiples of standard deviation. There are many possible designs for CUSUM control charts. The ARL curves shown in Figure 3.1 are commonly used and have good ability to detect shifts in the measurement system average:

Shift—the shift in the process average in standard deviation units.

AT&T—Shewhart with AT&T runs rules.

CUSUM—$H = 4.0$, $k = 0.5$.

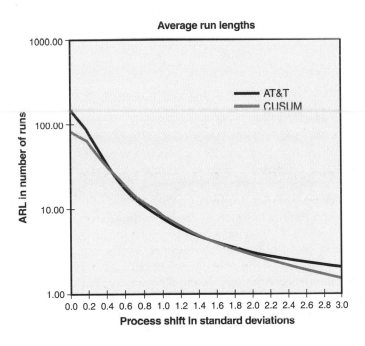

Figure 3.1   Average run lengths.

	ARL	
Shift	**AT&T**	**CUSUM**
0.0	91.75	167.7
0.2	66.80	93.42
0.4	36.61	38.58
0.6	20.90	19.45
0.8	13.25	11.94
1.0	9.22	8.38
1.2	6.89	6.42
1.4	5.41	5.20
1.6	4.41	4.37
1.8	3.68	3.78
2.0	3.13	3.34
2.2	2.70	3.00
2.4	2.35	2.73
2.6	2.07	2.52
2.8	1.85	2.34
3.0	1.67	2.19

For both charts, the average moving range can be used to design control charts for both the measurement system average and the measurement system variability.

The advantage of using the moving range is that it represents the variability of the measurement values used for the control chart decision. It includes both the short- and long-term components of the measurement variability. If the measurement system control sample size is two or more, a subgroup range chart can also be used to provide additional control of the within-subgroup variability.

Set up a periodic schedule to obtain the data for the measurement system control charts. Common schedules are once/shift or once/day. Design the control charts using the measurement results of the characterization study. In order to ensure good estimates of aim and standard deviation, the initial values should be recomputed after a total of 60 and 90 sampling intervals have been recorded, and adjusted if necessary. Recompute the within-subgroup range and/or the average moving range ($\overline{MR}$) after 30 or more measurement results have been recorded. The constants d_2, D_3, and D_4 are the standard control chart constants.[1]

Aim	Determined from the characterization study.
Standard Deviation(s)	Determined using measurement results from the characterization study. Using subgroup ranges: $s = \dfrac{\overline{R}}{d_2}$ Using average moving range: $s = \dfrac{\overline{MR}}{1.128}$, for $n = 2$

An alternative to estimating the standard deviation with the moving range involves the use of the mean square successive difference (MSSD) estimate.[39] Such estimates are referred to as SPROC and SPROCM for the production process and measurement system sources of variability, respectively.

Control Chart for Average

Control limits	Shewhart—Aim ± 1s, 2s, and 3s
	CUSUM—$H = 4.0s$, $K = 0.5s$

The design parameters for the standard CUSUM chart are h and k. These parameters are multiplied by the standard deviation and are denoted by the capital letters H and K.

The standard design parameters are:

$h = 4$ (analogous to 3 for Shewhart Charts)

$H = 4s$ (analogous to $3s$ for Shewhart Charts)

$k = 0.5$

$K = 0.5\,\Delta$

Where: Delta is the shift from aim you want to detect in a reasonable time. For standard CUSUM, Delta is equal to one standard deviation, thus $K = 0.5s$. Lucas[15-18, 33-38] offers a more comprehensive discussion of CUSUM charts, their design, parameter setting, and various refinements (for example, FIR). (See Section 5.3, "Example of CUSUM.")

Control Chart for Variability

Subgroup ranges—$D_4\,\overline{R}$, $D_3\,\overline{R}$

Average moving range—$D_4\,\overline{MR}$, $D_3\,\overline{MR}$

Following the schedule, plot the measurement results on the control chart for averages and the control chart for variability. If an out-of-control indication is given, follow the advice given in Section 3.7.1, "When Measurement System Is In-Control."[15–18] Remember AT&T run rules do not apply to moving range charts.

3.5.4 Blind Samples

In principle, and for measurement system control, _every sample should be handled exactly the same so that the measurement values can be trusted._ In reality, it is human nature to react to the labels on samples and treat some with more or less care. Laboratory standards and control samples tend to be handled with exceptional care. Rarely do control-sample control charts demonstrate the problems known to be common in the laboratory, such as sample loss and sample contamination, since the samples are known to be of special importance by the lab personnel and are treated with the utmost care.

It is also natural behavior for laboratory personnel to view blind samples as "unfair." In reality, blind samples are fair because they are handled in the same manner as routine product samples. One of the problems in keeping the blind sample truly blind is finding a source that can legitimately submit samples for analysis. A sample that appears every month from the Quality Assurance office is likely to be viewed after time with the same deference as a control sample. If, on the other hand, Quality Assurance maintains a set of blind samples which can be submitted at random by production personnel, the blind sample can serve a dual role of increasing cooperation and understanding between laboratory and production personnel while contributing to the control of the measurement system. The practice might also eliminate the deplorable use of production retains as substitutes for standards or control samples to assess measurement method control. Of all the practices used to make a short-term estimate of measurement method control, grab samples from retains are the most futile. If they are used, proper care must be exercised to ensure that:

- The retain is representative of the original sample;

- The retain has been stored under conditions that prevent change or deterioration; and

- Sufficient measurements are made to achieve a reasonable estimate of the desired properties level (usually $n > 15$).

If these steps are not followed, the use of the retained sample as a substitute for a standard or control sample can lead to misinformation.[4, 6–9, 22]

3.5.5 Controlling Measurement Systems on an Interlaboratory Basis

Any company that manufactures a product at multiple sites may find additional business advantages in controlling the measurement system on an interlaboratory basis. For example, transferring a product from one production site to another, or shipping the same product from two or more sites will be accomplished with more confidence knowing that the measurement results of key properties will be the same at all sites. Interlaboratory measurement results of product properties measured at the various laboratories should be equivalent, meaning that the data can be accepted by all parties, regardless of which laboratory did the measurement.

Many companies already do what is known as interlaboratory comparisons (ILC) within the company, or possibly with suppliers and/or customers. In ILC, the focus is usually to determine if multiple laboratories can measure a product at the same level for a given property. Sometimes, analysis of variance studies are included in the ILC, using a sampling plan that permits estimation of all of the important sources of variability in the particular measurement system. ILC always require statistical regression as discussed in Section 5.1 to compare the measurement results from the different laboratories. ILC may be done only once or a few times during the life of a product. Occasionally, ILC are done periodically, such as annually or biannually. Large biases discovered when analyzing data from ILC should trigger improvement activities at one or more of the laboratories, to reduce the bias to acceptable levels. Large sources of variability at any laboratory should trigger activities to reduce the variability. Well-planned ILC will produce data that can facilitate these improvement activities.

Recall from Section 3.5 that the intent of measurement system control is to provide timely information so that whenever the measurement system is not operating on target, laboratory personnel can find and fix the problem before measurement results reported to production process people have been seriously affected. Controlling on an interlaboratory basis may help diagnostic efforts, and it will be much easier to keep sites in agreement on a daily basis as well. Once established, ILC would simply become audits of how well the IMC program is doing. They would still add useful information about sources of variability within the measurement system.

Various names were used for these comparisons, including interlaboratory tests, interlaboratory studies, proficiency testing, measurement evaluation protocols, and round-robins. Most ILC are run across companies and/or independent laboratories, rather than within a single company, and most are done just one time to get the "answer." Use of ILC is evolving to where ILC are run within a company on a periodic basis.

ILC can also be designed to provide a great deal of information about within-laboratory measurement variability components, including operator variability, instrument variability, and variability over time. Analysis of variance (ANOVA) techniques are used to analyze the data produced by ILC, using a sampling plan that permits estimation of all the important sources of variability in the particular measurement system.

Some other objectives of ILC could be to determine:

- The robustness of the specific measurement system across testing personnel and/or instruments;

- The precision and sensitivity of an existing measurement system, using a particular measurement method;

- If the variability of the measurement system is consistent for different materials;

- If the measurement system can discriminate well enough among different levels of a property;

- If a measurement system can be improved by improving the measurement method;

- If the measurement system actually measures the property intended;

- The limits and fields within which the measurement system can be used;

- If there is correlation among measurement systems using different measurement methods for the same property;

- If the measurement system produces results that are consistent over time; and

- The usefulness of the measurement method in the C&PI for production process control and for product characterization and/or release decisions.

The specific objectives for ILC need to be carefully determined and incorporated into the measurement system. There are many standard practices for ILC for different industries; one compendium of these methods is published as *ASTM Standards on Precision and Bias for Various Applications*.[5]

It is beneficial to run two different ANOVA computations on the ILC data. The first should include laboratory as a variable to detect outliers and find any other anomalies within the data, and to make a statistical determination on whether one or more laboratories differs significantly from the others. This ANOVA is the source of the bias information. The second ANOVA proceeds as described earlier and is the source of within-laboratory variability information. Both ANOVAs use partially nested and crossed designs and are more complex than the ANOVAs for the laboratories separately. Appropriate software exists and ILC experts can provide the necessary help in running the correct ANOVAs. Final reports should include all biases and within-laboratory variances, as well as the variance components for each measured property and material at each laboratory, so that improvement efforts, if part of the ILC objective, can be placed in the most fruitful areas.[21–24, 26, 27, 29, 30]

3.5.6 Choosing What to Improve

Pareto analysis to prioritize the order in which measurement systems should be improved can be done based on several statistical analysis results. However, statistical analysis techniques do not take into account the business importance of improving a particular measurement system. The business importance of the production process and the customers' needs come first in setting priorities.

Pareto analysis of which measurement systems are least able to protect the production process will identify measurement systems that put the business at risk. This can be done by calculating how quickly a measurement system shift in average is detected by the measurement system control scheme, relative to how long the production process control scheme would take to detect that the measurement system had shifted. CUSUM is particularly suited for this task. (See Section 3.5.1, "Relative Sensitivity of Measurement and Production Control Charts.") Simply determine the average run lengths (ARLs from available tables) for a particular property for both production process control and for the corresponding measurement system control, based on the same sampling interval, for the production process to be off aim by the corresponding shift in the measurement results. Then assure that the time required to detect the shift for the measurement system control scheme is faster than the time required to detect that same shift by the production process control scheme. Twice as fast may be sufficient to allow corrective action on the measurement system before the production process CUSUM signals that a shift has occurred. In this way, the measurement system CUSUM control is protecting the production process from unnecessary adjustments or fruitless corrective action efforts. The measurement system is matched to the production system when the measurement system responds to measurement system change sufficiently faster than the production process detects the same measurement system change. For any comparison that yields an unacceptable measurement system to production process CUSUM match, something in the measurement system world needs to change. (See Section 5.6, "A New Measure for Measurement System Improvement Efforts" and 5.7, "Continually Updating Estimates of Measurement System Variance Components.")

3.6 OUTLIERS AND MISSING DATA ISSUES

As we have stressed so far, a measurement result must be considered valid in order to use the measurement result for its intended purpose. In the following discussion, we assume that the data being used in various calculations are valid. This means that the proper procedures were used to choose and measure the samples and that no obvious blunder occurred in generating the measurement result. In spite of all care, outliers will be generated as valid data and must be dealt with in some reasonable fashion. Over the years, authors seem to have avoided defining the term *outlier* in any rigorous way. Some early authors simply called an outlier "bad" data. To Natrella,[19] an outlier is a value that is "widely different from the rest" in a set of observations taken under the same conditions. In fact, something out of the ordinary usually occurs that causes the outlier, so there was a second underlying mechanism working to generate data. Our emphasis here will not be in deciding how to flag a value as an outlier, even though this step is crucial in order to make correct decisions. See Section 5.4, "Identifying an Outlier" for typical tests used to identify outliers.[20, 30]

The addition of electronic data collection and storage introduced several new mechanisms by which inappropriate or even incomprehensible data can be created. Velleman[42] identifies three categories:

1. "An *outlier* is any legitimate case whose value is far from the body of the data."

2. "A *blunder* is a clerical, digital, or measurement error in the data that is not informative about the real world (for example, a numerical datum that inappropriately contains alphabetical characters)."

3. "A *rogue* is correctly measured and recorded, but is inherently unusual. In no event should rogues be summarily discarded. A rogue can be worth more to your understanding of the data than all the ordinary cases."

Our interest lies in deciding what steps to take when an outlier has been flagged. What do I do next? The answer depends on the kind of data to which the outlier belongs.

One useful technique is to see if the decision would be different when using the measurement result versus when not using the measurement result. If the decision would be the same, keep the measurement result. If the decision would be different, some possible actions would be:

- Use the measurement result for decision making.

- Do not use the measurement result for decision making and wait for the next routine measurement result.

- Replace the measurement result by resampling or remeasuring.

- "Winsorize" the measurement result, that is, reduce the magnitude of the deviation, but retain the direction.

At times, eliminating an outlier from a set of data means that there will be missing data for some decision processes. Section 3.6.2 discusses some of the ramifications of missing data.

By following the suggestions in this section, outliers, which do exist in virtually all process industries to some extent, will not negatively impact the quality system and should be nearly eliminated over time. The cost savings, of course, will depend on the extent of the problems in the first place, but could be substantial or at least much larger than anyone would have thought. In any case, do not just remeasure or resample based on a reported measurement result that someone does not like. Do it so that improvements can be made, while protecting the customer.[28, 31, 32]

3.6.1 Outliers

The existence of outliers is frequently ignored or downplayed, but it isn't a good practice. If outliers are generated in the measurement system, several business costs are affected. Likewise, customers are affected if outliers represent production process or product problems. Reducing the number of outliers should be part of continuous improvement efforts. Sometimes outliers are not recorded, nor kept in any database, and are not the focus in a search for root causes. This makes it very easy to misunderstand the magnitude of the problem and the potential for improvement. It is not uncommon to find valuable information within the cause(s) of an outlier. Perhaps this is the greatest benefit from outliers. In a production environment, one typically finds four kinds of data: routine product, production process, measurement system control, and special studies. If an outlier occurs in any of these sources of data, there should be a mechanism in place to record the:

- Presence of an outlier;

- Outlier number itself;

- Suggestions for searching for root cause;

- The root causes for the outlier; and

- Corrective actions to prevent another occurrence of such an outlier.

Given this information, improvement projects can be implemented as needed to start reducing the size of the "hidden" laboratory and/or measurement system problems.

The time to decide what to do with an outlier is when it is flagged. What to do will be tackled for routine product data in the next section and for measurement system control data and special studies data in subsequent sections.

3.6.1.1 Response to Product Data Outliers

Questions that will have to be answered:

- Should I create a new measurement result using the same sample, if it is available? (Should I remeasure?)

- Should I take another sample and create a new measurement result? (Should I resample?)

- Should the measurement value be altered to a more innocent measurement value before proceeding with other calculations?

- Do I use the original measurement value in subsequent calculations and/or in making daily decisions?

- Should the original number be kept in the database as is?

- Should outlier information be shared with customers?

- How do I decide that it's time to seek improvement for a measurement system and/or measurement method?

When an outlier occurs in routine product data, there is one immediate question that should be answered. Did the outlier occur in the measurement system, or is it a real product problem? Failure to answer this may lead to a serious customer problem. In the following discussion, we will assume:

1. Some formal statistical analysis was performed on at least three months' worth of routine product data, preferably closer to a year's worth of data, to estimate the outlier standard deviation. This standard deviation has been used automatically for every sampling interval to decide if a measurement result is an outlier.

2. The associated product, whether in a sales unit form or not, will not be lowered from first-grade product simply because a single measurement result has been flagged as an outlier. Because a sales unit may be quite valuable or the environmental concerns may be too great, the product will be scrapped only as a last resort.

3. The production process will not be altered based on a single outlier, unless a search for root cause determines that the outlier is indeed a production process problem.

4. There is a strong desire within the production and support organizations to practice continual improvement in ways that bring about improved product quality.

Assuming that these four givens are in place, and that the suggestions found later in this section are followed, remeasuring or resampling should be a requirement. Many statisticians have preached against this practice because it was all too easy to remeasure/resample until an acceptable measurement result became available. The questionable measurement results (if in fact outliers were reported at all) were simply discarded and any information in them was lost. No record of their existence was available, since only the remeasure/resample measurement result was entered into the database. Product unacceptable to the customer may have been involved, but only a customer complaint would let this fact surface. This is, of course, a very poor quality practice.

The reason to remeasure/resample is to determine clearly whether the outlier cause was something abnormal in the measurement system world or something actually wrong with the production process. The time to search for root cause is at the time of occurrence. Waiting days or weeks before initiating a search for root cause will surely make the task much more difficult or impossible. When the outlier cause proves to have been in the measurement system, the same product unit or process sample could be remeasured enough times so that the product unit itself can be accepted as first-grade product.

Our suggestions are simple, at least in concept. When a measurement result is flagged as an outlier, do the following:

1. Remeasure or resample and measure the property again. Far more often than not, the new measurement result will not be an outlier. If in doubt whether the outlier came from the production process or measurement system, continue sampling and/or measuring until it is clear.

2. Do whatever is necessary to determine the extent of the problem whenever it has been proven that an outlier originated in the production process. Discard the sales unit associated with the outlier, or rework it if possible. Capture all sales units and product that might be a problem to customers. Continuous production processes pose especially difficult problems. The ultimate answer, of course, is to prevent such problems from occurring in the first place.

3. Put the new measurement result in the database as if it were the original measurement result.

4. Collect the remaining data for the current sampling interval, if any are needed.

5. Make all decisions using the new measurement result. For example, use the new measurement result for SPC purposes and shipping decisions.

6. Record the occurrence of the outlier somewhere and keep a count of the number of occurrences. Share this information with customers, in particular what corrective action was done because of the outlier, whenever a production process outlier has been identified.

7. Search for root cause immediately upon flagging the outlier and record the findings. Keep records on the types of root causes, whether from production process or measurement system, and institute improvement projects on those root causes that dominate. If the outlier was some kind of blunder, record this and record how the blunder happened. Learn to prevent all types of blunders.

Many "blunders" or "mistakes" happen and are discovered before a measurement result is reported. These are not outliers since no data was checked for the purpose of flagging an outlier. However, such events should be recorded, and retraining of operators/analysts, rewriting of the measurement method, or whatever action is needed should be initiated to prevent recurrence.

Note that doing the things suggested here means that outliers will not be in the database and will not affect continuous improvement measures. Recalculation of a new production process standard deviation should be easy and straightforward. Simple counts of various kinds, possibly plotted over time, may be sufficient measures to show improvement (or lack thereof). For example, plot the total number of process outliers and the total number of measurement outliers, by calendar quarter and by type of cause, in Pareto format. Things needing improvement are very likely to surface if the outliers are tracked as suggested.

3.6.1.2 Outliers in Measurement System Control Data

Dealing with measurement system control data outliers is much the same as dealing with out-of-calibration data. With measurement system control data, the time span since the measurement system was last known to be in-control should be quite short. If the outlier is caused by an out-of-control condition, there should be many fewer measurement results to evaluate for impact. If the cause is a CM supply that is actual product, the possibility of stored and shipped product having the same measurement results needs to be investigated.

The median absolute deviation (MAD) statistic (see Section 5.4, "Identifying an Outlier") can be used to easily estimate an outlier standard deviation at one moment in time. This can form the basis of an ongoing outlier test each time a new measurement system control result is generated. What is important here is whether or not the measurement system is off aim, and proper data must be used for making this decision. The same kinds of questions outlined in Section 3.6.1.1 need to be asked. The following show some of the additional actions:

1. Remeasure a sample from the same CM or a sample from another CM and measure the property again.

2. Determine if the measurement control material (CM) is defective in some way. Set a defective product unit aside and do not use it in the future for measurement system control purposes. If necessary, replace it or simply produce the measurement system control chart with one less product unit. Create a new measurement control supply, if no product unit is defined.

As in the case of production data, no outliers will actually be included in the database, so that whenever it is time to calculate a new measurement system standard deviation, the task should be easy and straightforward.

3.6.2 Missing Data—The Big Nuisance

The second data issue that must be dealt with is missing data. Sometimes missing production data occurs because one or more outliers have been eliminated from the data set. However, if the proper outlier and other procedures are in place and used religiously, missing data for daily decisions such as SPC and shipping product should not occur. At times it could be too expensive or untimely to choose a new sample (product unit or simple production process sample). It is probably easiest to make decisions on one less number than desired without adjusting the SPC system for the missing data. This should be a rare occurrence.

Doing the right things at the right time will ensure that missing product data for daily decisions is not an issue. Note that sufficient product data must exist in each sampling interval for proper daily decisions, which is used for SPC and shipping decisions. Otherwise, the product made during the interval having insufficient data should not be considered first grade. In effect, such product can be called nonconforming, since standard procedures have been violated.

Missing data presents a different problem whenever variance components are to be estimated or process standard deviations are calculated. Again, there should be few, if any, in the first place. If one or a few product data are missing, it may be sufficient to eliminate any incomplete data set and make all computations on the remaining data. There will normally be a large quantity of product data available, especially for sample sizes greater than one, the preferred situation. So, missing just a few sampling intervals of product data should not be a problem. For measurement system control data, these comments apply, with the understanding that there will usually be less data available than for product data analysis. But the primary reason to analyze measurement system control data is to estimate a meaningful measurement system standard deviation for measurement system SPC. For this purpose, missing just a few measurement sampling intervals should not present a problem, given that the data analysis is done on a quarterly basis, as recommended.

Missing data has the greatest consequence with maintenance data. There will normally be only one maintenance data set per day, because of the extra cost involved. There are numerous opportunities for creating mix-ups or for simply not getting the job done every day. Since these data are typically used only for estimating variance components, probably on a quarterly basis, the pressure to actually obtain all the required data every day is not there. So, it's reasonable to expect that more than a few days' data could be missing. Regardless, the recommendations have not changed. Ninety maintenance data sets are desirable, with 60 being a true minimum. Below 60, we suggest collecting maintenance data over a longer period of time before performing the analysis.

Following all the suggestions in this chapter should prevent the need for using advanced statistical analysis techniques that handle missing data and missing cells in such procedures as the analysis of variance. A few software packages exist that can perform such analyses, such as SAS and MINITAB.

3.7 PROPER RESPONSE TO A QUESTIONED MEASUREMENT RESULT

What should measurement personnel do when a customer questions the measurement result? First, determine whether the measurement system was in-control or out-of-control at the time of the measurement.

If the measurement system was in-control, the measurement result should be trusted. When a valid reason is found for questioning the mea surement result, this reason needs to be investigated. When the measurement system is in-control, an outlier measurement result means the product could be out-of-control. Measurement system control should minimize the frequency of measurement system outliers, allowing product outliers to be identified. See Section 3.6.1.1, "Response to Product Data Outliers" for further discussion. Questioning measurement results just because they are outside of some "limits" should be discouraged.

If the measurement system was out-of-control, the reported measurement result may be in error. The measurement result should not be used until the measurement system is corrected and validated. Then a new sam ple can be prepared and measured, and the new measurement result can be substituted for the questioned measurement result. This is further discussed in Section 3.6.1.2, "Outliers in Measurement System Control Data."

There should be a valid reason for questioning measurement results, such as "scientifically" impossible.

3.7.1 When Measurement System Is In-Control

When the measurement system is in-control, every measurement result should be trusted.

If the "limits" the measurement result is outside of have a valid statistical basis, the actions required to provide the planned protection should be taken. It does not matter if they are statistical process control limits for the production process, acceptance limits for incoming materials, or product release limits to protect against releasing product that does not meet specifications.

Sometimes a measurement value or result is so very different from other nearby measurement values or results that it is questioned. If the measurement system is in-control, the first action should be to determine if there was an assignable cause for the large deviation. The cause may be an error in the computation of the measurement result, an equipment failure, a contaminated sample, and so on. Depending on the assignable cause, the questioned measurement result should be replaced with a new measurement result either by remeasuring or resampling.

Resampling is the act of obtaining a completely new sample from the same material as that of the questioned measurement result. It is appropriate when replacing a measurement result for a sample that is known to be defective or invalid for an assignable cause. Remeasuring is the act of obtaining an additional measurement result on the same sample. It is appropriate when replacing a measurement result where the assignable cause is within the measurement system.

A measurement result outside of "limits" by itself is not a valid reason for resampling or remeasuring. Examples where resampling or remeasuring is appropriate:

- Incorrect identification of sample;

- Contaminated or damaged sample;

- Power failure during operation of measurement equipment; or

- Contaminated or expired reagent.

Resampling and/or remeasuring should *never* be used to simply replace measurement results that indicate nonconformity and, therefore, make nonconforming product into conforming product, that is, *purification by analysis*. If no assignable cause for a large deviation is identified, there are several procedures available to determine if a measurement result could be called an "outlier." (See Section 3.6, "Outliers and Missing Data Issues.")

Whatever choice you make should be based on your knowledge of the measurement system when the measurement result was generated, and the use of the measurement result. In any case, if a measurement result is replaced, the original measurement result should be retained and marked as replaced. The reason for remeasuring or replacing should be recorded.

3.7.2 When Measurement System Is Out-of-Control

First, it must be said that a measurement result from a measurement system that is out-of-control should never have been reported.

Second, if there was a measurement system control plan in place, no results should have been reported while the system was out-of-control. At the first out-of-control signal the laboratory should have stopped reporting measurement results.

Third, if there is no measurement system control plan in place, there is no valid response to a questioned measurement result. Get busy and implement a measurement system control plan so that measurement results can be trusted and not questioned.

3.8 LABORATORIES—WHAT IS IN THEIR NAMES?

Laboratories are labeled in many ways: Quality Control Lab, Advanced Technology Lab, Methods Development Lab, Research Lab, Customer Service Lab, Raw Material Lab, and so on. While the names are not important to this book, the functions performed in each are important.

The goal of a laboratory is often reflected in its name, that is, Quality Control Lab. The goal of a lab affects the mindset and actions of the laboratory personnel. Methods Development Lab personnel think about change and the creation of measurement methods. Quality Control Lab personnel think about control of the measurement system event, following the measurement method exactly, repeatability, and all of the other aspects of measurement systems that make them useable for business. Research Lab personnel think about what needs to be measured and what can be measured. Some of these goals conflict and can cause problems when mixed in one lab.

A person working in both the Methods Development Lab and Quality Control Lab could easily slip into being creative while doing a quality control measurement result, and might not be creative when needed while developing a measurement method. People tend to prefer one mind set; and when rushed, not be as attentive to work requiring another mind set.

While utopia would be a physically separate laboratory with different personnel for each different and conflicting goal, the real world is not so generous. People are asked to achieve several, and sometimes all, of the goals of these different laboratories. The key is for personnel to be aware of the pitfalls and to have procedures that provide checks and balances to help keep the person in the correct mind set for each measurement result made. Variability and bias are natural parts of human nature. It is, therefore, prudent to establish checklists, record requirements and other things that remind people of what they are doing, and keep records of what is done so that deviations can be found quickly. It may also be possible to have an external laboratory do work that is needed infrequently. An independent laboratory could be contracted to develop measurement methods. Measurement methods developed by industry standards associations can also be used.

The quality of a laboratory measurement result can be significantly impacted by the number of different goals of the laboratory and the conflicts between them. Accept the situation and deal with it proactively, or you will deal with the problems that it causes.

3.9 THE RESEARCH ENVIRONMENT

Research is about finding something new by exploration and conducting experiments that test the null hypothesis, that is, that nothing is new. Research is the area above all others where this book has strategic importance. In research, decisions must be made with quantified uncertainty in the measurement results. This is an important place to know we can trust our measurement results. Researchers fear measurement system rules will inhibit their creativity and discoveries but, in fact, the rules separate true discoveries from measurement system variability. Measurement system rules permit you to know when you have succeeded. Do you throw away or patent an outlier? (See Section 3.6, "Outliers and Missing Data Issues.")

For companies working to keep ahead of their competition by creating new products, research measurement results are strategic knowledge. Research measurement results can change the company direction and trigger major expenditures, both of which determine the future of a company. With a whole company's future at stake, the measurement results must be trustworthy.

Trusted measurement results are the most essential component of trusted research conclusions. These require a trusted measurement system that adheres to all the principles you have read in Chapter 2 and Chapter 3 of this book.

Being trustworthy does not mean a measurement system will not have uncertainty. It means that the uncertainty will be quantified. You will know how much trust you can place in the measurement results.

3.10 REFERENCES

1. Douglas C. Montgomery, *Introduction to Statistical Quality Control*, Second edition (New York: John Wiley & Sons, 1991): 391–396.
2. AIAG, *Measurement Systems Analysis—Reference Manual* (Southfield, MI: Automotive Industry Action Group, 1990).
3. ASQC Chemical and Process Industries Division, Chemical Interest Committee, *Specifications for the Chemical and Process Industries—A Manual for Development and Use* (Milwaukee: ASQC Quality Press, 1996).
4. John Mandel, *Evaluation and Control of Measurements* (New York: Marcel Dekker, 1991).
5. J. M. Juran, *Quality Handbook*, Fifth edition (New York: McGraw-Hill, 1998), chapter 27.
6. James P. Dux, *Handbook of Quality Assurance for the Analytical Chemistry Laboratory*, Second edition (New York: Von Nostrand, 1990).

7. International Organization for Standardization, *ISO 5725-1 International Standard, "Accuracy (Trueness and Precision) of Measurement Methods and Results, Part 1: General Principles and Definitions* (Geneva, Switzerland: International Organization for Standardization, 1994).

8. International Organization for Standardization, *ISO 5722-3 International Standard, Accuracy (Trueness and Precision) of Measurement Methods and Results Part 3: Intermediate Measures of the Precision of a Standard Measurement Method* (Geneva, Switzerland: International Organization for Standardization, 1994).

9. Robert W. Berger and Thomas H. Hart, *Statistical Process Control* (New York: Marcel Dekker, 1986).

10. Ford Motor Company, *Q101 Quality System Standard* (Detroit, MI: Ford Motor Company, 1986).

11. AT&T Technologies, *Statistical Quality Control Handbook* (Indianapolis, IN: AT&T Technologies, 1984).

12. Ellis Ott, *Process Quality Control* (New York: McGraw Hill, 2000).

13. Paul F. Velleman, *DataDesk Plus Handbook* (Ithaca, NY: Data Description Inc., 1996).

14. N. I. Johnson and F. C. Leone, *Cumulative Sum Control Charts* (Industrial Quality Control, June, July, August 1962).

15. J. M. Lucas, "A Modified V-Mask Control Scheme," *Technometrics* 15 (1973): 833–847.

16. J. M. Lucas, "The Design and Use of V-Mask Control Schemes," *Journal of Quality Technology* 8 (1976): 1–12.

17. J. M. Lucas, "Combined Shewhart-CUSUM Quality Control Schemes," *Journal of Quality Technology* 14 (1982): 51–59.

18. J. M. Lucas and R. B. Crosier, "Fast Initial Response (FIR) for CUSUM Quality Control Schemes: Give Your CUSUM a Head Start," *Technometrics* 24, (1982): 199–205.

19. M. G. Natrella, *Experimental Statistics—Handbook 91* (Washington D.C.: National Bureau of Standards, 1963).

20. W. J. Dixon and F. J. Massey, Jr., *Introduction to Statistical Analysis*, Third Edition (New York: McGraw-Hill, 1969): 328–332.

References 21 through 28 are from American Society for Testing of Materials (ASTM), West Conshohocken, PA

21. ASTM, D1749-93 (1997) *Standard Practice for Interlaboratory Evaluation of Test Methods Used with Paper and Paper Products, Annex A1*, (ASTM).

22. ASTM, BIAS97 (1997) *Standards on Precision and Bias for Various Applications, ASTM, Fifth Edition*, (ASTM).

23. ASTM, D4210-89 (1996) *Standard Practice for Intralaboratory Quality Control Procedures and a Discussion on Reporting Low-level Data*, (ASTM).

24. ASTM, D4467-94 *Standard Practice for Interlaboratory Testing of a Test Method that Produced Non-normally Distributed Data*, (ASTM).

25. ASTM, E826-85 (1996) *Standard Practice for Testing Homogeneity of Materials for the Development of Reference Materials*, (ASTM).

26. ASTM, E691-99 *Standard Practice for Conducting an Interlaboratory Study to Determine the Precision of a Test Method*, (ASTM).

27. ASTM, E601-98 *Standard Practice for Conducting an Intralaboratory Study to Evaluate the Performance of an Analytical Method*, (ASTM).

28. ASTM, E178-94 *Standard Practice for Dealing with Outlying Observations*, (ASTM).

29. Donald J. Wheeler and Richard W. Lyday, *Evaluating the Measurement Process*, Second edition (Knoxville, TN: SPC Press, 1989): 49–60.

30. C. R. Hicks, *Fundamental Concepts in the Design of Experiments* (New York: Holt, Rinehart, and Winston, 1964).

31. Boris Inglewicz and David C. Hoaglin, *Volume 16: How to Detect and Handle Outliers* (Milwaukee: ASQC Quality Press, 1993).

32. ASQC Statistics Division, *Glossary and Tables for Statistical Quality Control*, Third edition (Milwaukee: ASQC Quality Press, 1996): item 6.16 and 6.17.

33. J. M. Lucas, "Combined Shewhart-CUSUM Quality Control Schemes," *Journal of Quality Technology* 14, (1982): 51–59.

34. J. M. Lucas and R. B. Crosier, "Robust CUSUM: A Robustness Study for CUSUM Quality Control Schemes," *Communications in Statistics—Theory and Methods* 11 (1982): 2669–2687.

35. J. M. Lucas, "Counted Data CUSUMs," *Technometrics* 27 (1985): 129–144.

36. J. M. Lucas, "Cumulative Sum (CUSUM) Control Schemes," *Communications in Statistics—Theory and Methods* 14 (1985): 2689–2704.

37. J. M. Lucas and K. K. Hockman, "Variability Reduction through Subvessel CUSUM Control," *Journal of Quality Technology* 19 (1987): 113–121.

38. J. M. Lucas, "Control Schemes for Low Count Levels," *Journal of Quality Technology* 21 (1989): 199–201.

39. D. W. Marquardt, "Estimating the Standard Deviation for Statistical Process Control," *International Journal of Quality and Reliability Management* 10, no. 8 (1993): 60–67.

40. Donald Fels, "Fuzzy Math: A Count Too Close to Ever Be Clear," *San Francisco Chronicle* 27 (November, 2000).

4

Insights on Future Trends in Measurement

The general trend is away from central laboratories, less finished product measurement, and increased usage of sensors, instruments, and online measurements.

4.1 PRODUCTION AREA MEASUREMENTS BY OPERATORS

In an attempt to save time, reduce costs and measurement laboratory personnel, and provide timely data to the decision maker, many companies in the C&PI are moving the product or production process measurement functions down to the decision maker—the production operator. Production line operators are taking samples and running the required measurements at some point in the production process. In so doing, an interesting scenario has developed: Can a production line operator make an independent decision on product disposition based on the measurement results of their own measurements? Have we created a conflict of interest? If the operator's measurement results for a product property are within the specification range, but the operator knows the production process measures have not remained in-control, can that operator judge the product to be good? Perhaps the measurement result is outside of the specification limit and production process variables are in-control. Is the product good or are we being influenced by a poor measurement result? Will quality or productivity rule, or will companies use the combination of timely product measurement results with production process variables measurement results as means to increase knowledge upon which better decisions can be made? What are the real risks?

4.1.1 Operator Training

Answers to some of the questions just posed depend on the measurement results. In the case of production area measuring by operators, one of the major components that drive the correct decision is the ability of the operator to properly perform the required measurement procedure. This will depend on the quality of training that production operators are given, the content of the training, and how they are certified as qualified testers. Of course, we all realize that the culture of the company and the roles of quality and productivity in driving performance excellence are key influencing factors.

It is important to understand that the training of production operators to generate measurement results for material properties and make informed decisions about disposition should be affected by all the topics of this book, particularly Chapter 2 and Chapter 3. Once again we ask, "How can I trust this measurement result?" The answer lies in the understanding of the content of this book and the training that operators receive. Remember that the operator is now a major component of the measurement system.

4.1.2 Production Area Measurement Systems

Before production area measurement systems, operators made decisions based on the measurement results from the offline measurement laboratory. Operators did not do calibrations, take part in interlaboratory comparisons (ILC), run measurement methods, and so on. Now the operators are a major component in the measurement system.

Let's assume the current company measurement laboratories are being essentially replaced by production area measurement systems for all product lines. In addition to acquiring the knowledge and experience in running measurement methods on the final product, the operators could be:

- Implementing SPC on their measurement results;

- Evaluating the measurement system with SPC;

- Participating in ILC studies;

- Calibrating measurement equipment;

- Maintaining measurement systems that are included in an ISO 9000 registration;

- Solving measurement system problems; and

- Participating in R&R studies.

Are you getting the feeling that the contents of this book up to this section form the body of knowledge, training requirements, understanding, and involvement for production operators? You should be. In fact, the transfer of the measurement function, no matter how small, from the measurement laboratories to the production operators is not just a simple task of training these operators how to run the necessary measurement methods. It involves the transfer of a measurement system to the production floor. The extent of the transfer dictates what has to take place if the operators, our new testers, are to make informed decisions about the production process and the product it produces based on facts gleaned from available data. They may also be involved in measurement system assurance. Using the contents of this book as a guide will help insure the integrity of the production area measurement system.

4.2 INLINE MEASUREMENTS REPLACING OFFLINE MEASUREMENTS

As sensors and other instrument technologies develop, it will become easier to reliably measure upstream production process variables and other in-process product properties. In fact in many industries, technology has provided this opportunity. Inline measurement results are obtained from devices in contact with the production process stream. Inline measurement results are replacing offline measurement results as the basis for day-to-day decisions regarding product quality. This changes the role of the current measurement or control laboratories to be more oriented to measurement system assurance rather than measurement system control. Perhaps the major tasks will become one of assuring that these in-line measurement system devices are giving us data we can trust, and therefore use to make informed decisions.

4.2.1 Production Process Variables versus Product Properties

As production sites use more process control systems and instrumentation, the decision criteria become an important piece of production process information. Pressures, temperatures, flows, and so on, as well as inline product measurement results, are the actual controls of the production process. These measurement results are influenced and automatically controlled by feedback control loops.

Moving upstream increases our production process knowledge by understanding the relationships between product characteristics (Y's) and production process variables (X_i's), $Y = f(X_i)$. In developing this level of understanding, we will also learn how to control the production process variables and inline product measurement results in an attempt to minimize their impact on the final product characteristics. However, the issues don't change. We still must have measurement values and measurement results we can trust. That brings us back to the beginning, Chapter 2, of this book.

As more and more automation is introduced into our production processes, the upstream measurement results, measurement methods, sensors, suppliers of sensors, and so on, become the components of the new inline measurement system and its sources of variation. As such, everything discussed about the measurement system in this book applies to these new measurement systems.

We will still need a measurement function as part of the organization, but the role will shift to assuring the reliability of the inline measurement devices and sensors. This will assure confidence in the measurement results, upon which we can make sound decisions regarding production process behavior and, ultimately, product disposition.

4.2.2 Product Disposition and Inline Measurement Results

Another movement that changes the role of the measurement or control laboratories is a direct outcome of technology. As sensor development progresses more and more, companies are implementing what could be called "ship by gage." The product property measurement results from automated sensors are used to determine disposition of the product. Of course, in many of the current measurement laboratories, automated measuring equipment is used to produce the quality control measurement results. This concept of product disposition is not new to many companies, but the change involves the placement of the sensor. The sensors are becoming an integral part of the production process and measurement system.

What changes? None of the principles and practices in this book change. We still need to:

- Sample the production process;

- Calibrate the measurement equipment;

- Understand the measurement system;

- Control the measurement system; and

- Trust the measurement vales and measurement results.

The role of the control or measurement laboratories remains the same with respect to focus, but changes exist in the implementation of this role. For example, some instruments or sensors have an automated calibration procedure. We need to understand these procedures, develop trust in their measurement results, set up methods to verify the calibration, and develop the necessary maintenance methods for sensors—this is measurement assurance. Understanding the contents of this book will prepare companies for the implementation of product disposition by inline instrument measurement system.

4.3 MEASUREMENT RESULTS AND DECISION LIMITS IN THE TWENTY-FIRST CENTURY

Toward the end of the twentieth century, thinking in the measurement community changed drastically on the subject of decision rules. The issue came down to this: Now that we have a standardized method of determining and expressing measurement uncertainty for most measurement processes, we need to have more uniform ways of respecting the uncertainty of measurement when writing and carrying out decision rules.

The issue is that there are measurement results that do not strictly define "pass" or "fail" because they are close enough to a decision threshold that the uncertainty throws the decision into question. In the end, most standards bodies are recommending a third choice, so that we will have "pass," "fail," or "uncertain." This simple difference is such a large change in the way that we think about decisions that it is sure to engender controversy for decades to come. These "uncertain" measurement results put the issue of producer versus consumer risk squarely on the negotiation table.

Producers and consumers view the risks associated with these "uncertain" measurement results differently. A consumer might add "guard bands." In the simplest case, anything that fell into the "uncertain" category would just report as "fail." This is safe (because it takes no chances of passing defective material), but might be expensive (because it rejects possibly good material). Where measurement uncertainty is large, guard bands could easily make the difference between success and failure of a production process.

Two areas in which standards bodies have addressed this issue are:

1. ISO/IEC 17025 states in paragraph 5.10.4.2: *"The calibration certificate shall relate only to quantities and the results of functional tests. If a statement of compliance with a specification is made, this shall identify which clauses of the specification are met or not met. When a statement of compliance with a specification is made omitting the measurement results and associated uncertainties, the laboratory shall record those results and maintain them for possible future reference. When statements of compliance are made, the uncertainty of measurement shall be taken into account."*

A "statement of compliance with a specification" includes statements of "pass" or "fail" when a specification contains a decision rule such as presented in Section 3.2. The "uncertainty of measurement shall be taken into account" can be interpreted to require either statement of an uncertain condition or use of a guard band so that uncertainty is explicitly considered in the application of decision rules. Many organizations and standards-enforcing bodies are interpreting this requirement in exactly that way.

2. ISO/IEC 14253-1, *Geometrical Product Specifications— Inspection by Measurement of Work Pieces and Measuring Instruments, Part 1. Decision rules for proving conformance or nonconformance with specifications* was written in the context of dimensional measurements. But the rules explicitly require an "uncertain zone" as well as a "conformance zone" and a "nonconformance zone" when interpreting measurement results in the context of uncertainty for the purpose of determining conformance to specifications.

It is best to avoid writing specifications that require decisions be made with "uncertain" measurement results. If such specifications are received from a customer, the supplier should work with the customer to improve the measurement capability or write a new specification that does not require "uncertain" measurement results be used to make pass/fail decisions while protecting both the customer and supplier.

4.4 "END STATE"

By way of the title of this section, one could surmise that the end of the book is near. It is, but the "end" referred to here is the "end state" that this book has taken us to objectively. One can step back and review where we have come from, where the book has taken us and, finally, ask, Where are we going from here?

Some readers might have come to this book with little knowledge about measurement systems. Others might have desired to increase their knowledge. In any case, the goal is to increase our knowledge and apply it in our jobs in order to help our organizations improve. If the material in this book is followed, the outcome will be trusted measurement results on which to base our decisions pertaining to the purpose of generating the measurement results in the first place. These decisions should enable companies to satisfy customers and better understand how to improve processes. Combine this with the impact of technology, automation, automated sensors, and measurement methods, and one can begin to see what the future holds for measurement systems—the "end state."

The natural progression from application of the concepts in this book would be controlled production processes that produce products "so good" that finished product measurement is unnecessary. Quality control laboratories, measurement laboratories, and production operator measurements would be eliminated. Could this be the "end state?" It could be. If so:

- Would we have to do any type of measuring?

- Would most of the measuring, as we know it today, be transformed to an auditing function?

- Would new measurement systems and methods evolve as products and technology evolves? Would this keep this book a living document?

- Can we ever reach the "end state" because of our corporate cultures and government regulations?

These questions and others become the questions that must be raised when we reach the "end state." The answers to such questions would define the "end state." Sounds like we have come full circle, but in doing so, one can only hope that somewhere along the way, measurement systems have improved and better decisions are being made based on better measurement results.

5

Appendix

5.1 COMPARING TWO MEASUREMENT METHODS

As mentioned in Sections 2.3, 3.4, and 3.7, statistical regression is needed when comparing two measurement methods, A and B. However, it is important to note that standard least squares regression is not appropriate for comparing measurement methods. Standard regression techniques assume that the "X" variable is known and without error. When comparing two measurement methods, both variables "X" (method A measurement results) and "Y" (method B measurement results) are subject to error. Two useful statistical analysis techniques to handle this situation that give comparable results are those in Mandel[1] and Davies.[2] The computational procedure is given for computing both the slope and its standard error. Although computational procedures are given for the case when the error variances are unknown, it is good practice to obtain estimates of the error variances for the two measurement methods.

Jackson[3] outlines a procedure that computes the orthogonal regression line between two related variables. This procedure also has additional uses beyond the comparison of two measurement methods.

Another useful statistical analysis technique is to use the data from the parallel sampling experiment for the two measurement methods and estimate the variance components for samples and measurement error. This procedure is also useful for comparing three or more measurement instruments or measurement sites.

For example, variance components from ANOVA:

	Method X	Method Y
Sample to sample	$s^2_{samp}(X)$	$s^2_{samp}(Y)$
Within sample (measurement)	$s^2_{meas}(X)$	$s^2_{meas}(Y)$

Assuming no variation in X and a linear relationship (Y= a + bX), the slope and intercept are determined by:

$$b = \sqrt{\frac{s^2_{samp}(Y)}{s^2_{samp}(X)}} \qquad a = Yavg - b \cdot Xavg$$

These procedures are compared using a common set of data.
For example:

- 30 samples.

- Two measurements by each measurement method for each sample.

- The two measurement values for methods A and B were averaged and are shown in the following table as $\overline{A}$ and $\overline{B}$.

The data are listed and summarized below.

Sample	A(1)	A(2)	B(1)	B(2)		$\overline{A}$	$\overline{B}$
1	87	89	130	131		88.0	130.5
2	100	103	157	157		101.5	157.0
3	102	105	152	156		103.5	154.0
4	116	117	169	166		116.5	167.5
5	89	94	142	142		91.5	142.0
6	95	97	150	150		96.0	150.0
7	107	96	149	151		101.5	150.0
8	113	117	158	160		115.0	159.0
9	107	112	160	161		109.5	160.5
10	101	97	141	140		99.0	140.5
11	113	108	164	167		110.5	165.5
12	98	99	157	156		98.5	156.5
13	96	97	147	142		96.5	144.5
14	91	93	144	142		92.0	143.0
15	89	93	131	133		91.0	132.0
16	91	93	139	135		92.0	137.0
17	95	100	146	143		97.5	144.5

(continued)

Sample	A(1)	A(2)	B(1)	B(2)		$\overline{A}$	$\overline{B}$
18	99	97	148	155		98.0	151.5
19	110	108	165	162		109.0	163.5
20	91	89	132	136		90.0	134.0
21	81	89	120	118		85.0	119.0
22	96	97	146	151		96.5	148.5
23	81	85	125	124		93.0	124.5
24	99	100	153	150		99.5	151.5
25	104	101	154	152		102.5	153.0
26	96	96	151	147		96.0	149.0
27	118	108	165	160		113.0	162.5
28	105	96	147	153		100.5	150.0
29	91	91	146	143		91.0	144.5
30	93	98	147	143		95.5	145.0
					MEAN	98.65	147.68

5.1.1 Variance Component Analysis

The variance analysis for each measurement method is shown below.

Method A

Source	df	SS	MS	F	EMS	Variance component
Samples	29	4245.1	146.693	14.3	$\sigma_e^2 + 2\sigma_s^2$	68.222
Measurement	30	307.5	10.250		σ_e^2	10.250

$$s^2_{\text{samp}}(A) = 68.222$$

$$s^2_{\text{meas}}(A) = 10.250$$

$$s^2\,\overline{A} = \frac{s_e^2}{n} = \frac{10.250}{2} = 5.125$$

Method B

Source	df	SS	MS	F	EMS	Variance component
Samples	29	8197.5	282.672	52.5	$\sigma_e^2 + 2\sigma_s^2$	138.645
Measurement	30	161.5	5.383		σ_e^2	5.383

$$s^2_{samp} (B) = 138.645$$

$$s^2_{meas} (B) = 5.383$$

$$s^2 \overline{B} = \frac{s_e^2}{n} = \frac{5.383}{2} = 2.692$$

$$b = \sqrt{\frac{s^2_{samp} (B)}{s^2_{samp} (A)}} \qquad a = Bavg - b \bullet Aavg$$

$$b = \sqrt{\frac{138.645}{68.222}} = 1.426$$

$$a = 147.6833 - (1.426)(98.65) = 7.008$$

$$Method\ Y = 7.008 + 1.426 \bullet Method\ X$$

In all cases, when comparing two (or more) methods, it is desirable to determine the relative slopes of the two lines with the error component "removed." The variance component method determines the "slope" of one method relative to another by the ratio of the sample-to-sample standard deviations as determined from the variance components of the same samples measured by each method. The sample-to-sample components in standard deviation form have the measurement error removed.

5.1.2 Regression Analysis (Both Variables Subject to Error)

In addition to the data for the two measurement methods, the ratio of the two error variances is required. Usually, repeat measurement results are obtained when initially collecting the data.

Example

$$\frac{s^2_{\overline{Y}}}{s^2_{\overline{X}}} = \frac{5.125}{2.692} = 1.90$$

Both Davies'[2] and Mandel's[1] procedures give essentially the same numbers for a and b.

	a	b	Std error (b)
Davies (Errors known)	3.37	1.463	.186
Mandel (Errors known)	3.37	1.463	.118

The following table is a comparison of the variance component to the procedures of Davies and Mandel. Shown is the $Y = a + bX$ relationship over the range included in the sampling study.

	V Comp	Davies	Mandel
Constant (a)	7.01	3.37	3.37
Slope (b)	1.426	1.463	1.463

Method A	Method B		
75	114	113	113
80	121	120	120
85	128	128	128
90	135	135	135
95	142	142	142
100	150	150	150
105	157	157	157
110	164	164	164
115	171	172	172

The three procedures give essentially equivalent results.

5.1.3 Using the Result

1. Determine the relative signal-to-noise ratios.

   The signal-to-noise ratio is the ratio of the sample-to-sample standard deviation to the measurement standard deviation. It's a measure of relative sensitivity of a measurement method:

   $$\frac{\text{Signal}}{\text{Noise}} = \frac{s^2_{samp}}{s^2_{meas}}$$

   $$\text{Method X:} \quad \sqrt{\frac{68.222}{10.25}} = 2.59$$

   $$\text{Method Y:} \quad \sqrt{\frac{138.645}{5.383}} = 5.08$$

   Thus measurement method B is much more sensitive than measurement method A.

2. Use the relationship $Y = a + bX$ to update targets, specifications, control charts, and acceptance/release plans.

3. Relate specification limits using measurement method B to those using measurement method A.

   If the specifications are on true measurement results as recommended in *Specifications in the Chemical and Process Industries,*[4] the specifications are related by the relationship $Y = a + bX$.

   Specification B = a + b (Specification A)

   For example, determine specification limits for method B from limits for method A and the relationship determined by the variance component analysis.

   (B = 7.01 + 1.426 A)

Limits	Method A	Method B
Low	85	128.2
Target	95	142.5
High	105	156.7

4. Acceptance plans

   Acceptance plans using measurement method B in place of measurement method A should take into account both the newly developed specification limits and the change in measurement method precision.

5. Continual improvement

   If the sampling experiment is designed to estimate both long- and short-term measurement errors, the analysis of variance result may indicate areas for future technical effort to improve the measurement method precision. A more detailed sampling study involving operators, equipment, and procedures can be designed to identify specific elements of the measurement method for improvement.

REFERENCES

1. J. Mandel, "Fitting Straight Lines When Both Variables Are Subject to Error," *Journal of Quality Technology* 16, no. 1 (January 1984).
2. O. L. Davies, *Statistical Methods for Research and Production* (New York: Hafner Publishing, 1957): Section 7.6.
3. J. E. Jackson, "Quality Control Methods for Two Related Variables," *Industrial Quality Control* 12, no. 7 (January 1956).
4. ASQC Chemical and Process Industries Division, Chemical Interest Committee, *Specifications for the Chemical and Process Industries—A Manual for Development and Use* (Milwaukee: ASQC Quality Press, 1996).

5.2 CALIBRATION

There is variability in all measurement systems and materials. Even a standard reference material (SRM) has variability, albeit small. A standards laboratory should report the standardized uncertainty of measurement values for an SRM. Too often, many people take standards laboratory measurement values as absolute truth. Calibration in the C&PI is the comparison of the measurement values from two different measurement systems using an SRM that has variability. This is a process with uncertainty and must be approached with caution, as discussed in Section 2.3, "Calibration and Uncertainty."

Because of variability, we need the application of statistical regression to chemical calibration, which is more difficult than it first appears because the standards laboratory measurement values have uncertainty. Ordinary regression assumes that the independent variable x is known with high precision and the dependent variable y with low precision relative to x. Computation by least squares yields the usual estimates for the slope and intercept of a straight line. This is typically used to predict values of y from various values of x. Chemical calibration is more complex in that values of y are used to predict values of x.[1-7] Furthermore, while it is easy to compute an x for a measured y, "for a measurement value to have any interpretable merit, an estimate of its uncertainty must be reported."[1]

There are five fundamental assumptions (from a statistical point of view) in calibration.[1] They are: (1) a linear model; (2) error only in Y; (3) random and homogeneous error; (4) uncorrelated errors; and (5) normally distributed error. In the usual straight-line model, the intercept is often needed, for example, because of a blank reading in the standards. Since both a slope and an intercept are computed this is a bivariate problem,[4] as

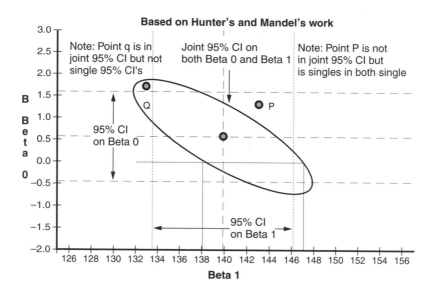

Figure 5.1   Calibration uncertainty.

shown in Figure 5.1. That is, it is not just a simple matter of computing the marginal *t* distributions on the slope and intercept.

The uncertainty in the calibration line is computed based on the Working-Hotelling[9] confidence region,

$$b_0 + b_1 \pm \sqrt{2 \times F_{2, v, a} \times [\frac{1}{n} + \frac{(x - x)^2}{S_x^2}] \times s^2}$$

Note that the *F* allows for 2 and *v* degrees of freedom so that it can be applied over the entire range of *x* and not just at x_0. Hunter[3] says, "The reader should note that the *t* interval is valid only for a single predicted mean response at some preselected value x_0 whereas the Working-Hotelling-Scheffe interval, which employs Snedecor's *F*, is valid whatever the setting of *x*." Many otherwise good literature references mistakenly use *t* instead of *F*. Many statistical books such as Draper and Smith[7] analyze the complex calibration problem correctly, using *F* instead of *t*.

Figure 5.2 shows how all this translates to a final uncertainty in the measured *x*. Other considerations in calibration are where to place the *x* values, cost, and so on.

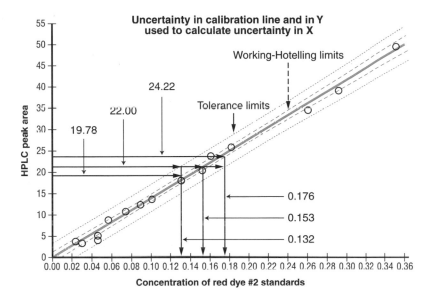

Figure 5.2 Calculation of calibration uncertainty.

In addition to statistical problems, there are chemical problems leading to systematic errors, such as: (1) recovery, (2) interferences, (3) selectivity, (4) range, and (5) sensitivity.[10-12]

REFERENCES

1. Douglas L. MacTaggart and Sherry O. Farwell, "Analytical Use of Linear Regression, Part I: Regression Procedures for Calibration and Quantitation," *Journal of the Association of Official Analytical Chemists* 75, no. 4 (1992): 594–608.
2. Christine Osborne, "Statistical Calibration: A Review," *International Statistical Review* 59 (1991): 309–336.
3. J. Stuart Hunter, "Calibration and the Straight Line: Current Statistical Practices," *Journal of the Association of Official Analytical Chemists* 64, no. 3 (1981): 574–583.
4. John Mandel and Frederic J. Linnig, "Study of Accuracy in Chemical Analysis Using Linear Calibration Curves," *Analytical Chemistry* 29, no. 5 (May 1957): 743–749.

5. James N. Miller, "Basic Statistical Methods for Analytical Chemistry, Part 2: Calibration and Regression Methods, a Review," *Analyst* 116 (January 1991): 3–14.

6. Catherine J. Bailey, Elizabeth A. Cox, and Janet A. Springer, "High Pressure Liquid Chromatographic Determination of the Intermediate/Side Reaction Products in FD&C Red No. 2 and FD&C Yellow No. 5: Statistical Analysis of Instrument Response," *Journal of the Association of Official Analytical Chemists* 61, no. 6 (1978): 1404–1416.

7. N. R. Draper and H. Smith, *Applied Regression Analysis*, Second edition (New York: John Wiley & Sons, 1981).

8. H. Scheffe, *The Analysis of Variance* (New York: John Wiley & Sons, 1959).

9. H. Working and H. Hotelling, *Journal of American Statistical Association*, Supplement (Proc.) 24 (1929): 73–85.

10. W. Funk, V. Dammann, and G. Donnevert, *Quality Assurance in Analytical Chemistry* (New York: VCH Publishers, 1995).

11. I. R. Juniper, "Method Validation: An Essential Element in Quality Assurance," in *Quality Assurance and TQM for Analytical Laboratories*. ed. M. Parkany (Royal Society of Chemistry, 1995).

12. Peter C. Meier and Richard E. Zund, *Statistical Methods in Analytical Chemistry* (New York: John Wiley & Sons, 1993).

5.3 EXAMPLE OF CUSUM

See Section 3.5.2 and Section 3.5.3 for definitions of the terms used in this example of using the CUSUM technique.

Aim = 3.20

s = 0.09

Delta = 0.09

h = 4.0

k = 0.5

$H = h \times s = 4.0(0.09) = 0.36$

$K = k \times s = 0.5(0.09) = 0.045$

The data in sequence

3.38	3.32	3.13	3.25	3.22	3.09	3.13	3.16	3.30	3.29	2.99	3.18
3.14	3.13	3.05	3.11	3.13	3.21	3.20	2.97	3.16	3.06	3.15	3.14

For the first data point, SL and SH are set to zero:

SH = Max [0, (X(i) – Aim) – K + SH (i – 1)]

  = Max [0, (3.38 – 3.20) – 0.045 + 0] = 0.135

NH = 1

SL = Max [0, -(X(i) – Aim) – K + SL (i – 1)]

NL = Max [0. – (3.38–3.20) – 0.045 + 0] = Max [0, -0.225] = 0

NL = 0

For the second data point:

SH = Max [0, (X(i) – Aim) – K + SH (i – 1)]

  = Max [0, (3.32 – 3.20) – 0.045 + 0.135] = 0.210

NH = 2

SL = Max [0, -(X(i) – Aim) – K + SL (i – 1)]

  = Max [0, -(3.32 – 3.20) – 0.045 + 0] = Max [0, -0.165] = 0

NL = 0

For the third data point:

SH = Max [0, (X(i) – Aim) – K + SH (i – 1)]

  = Max [0, (3.13 – 3.20) – 0.045 + 0.210] = 0.095

NH = 3

SL = Max [0, -(X(i) – Aim) – K + SL (i – 1)]

  = Max [0, -(3.13 – 3.20) – 0.045 + 0] = Max [0, 0.02] = 0.025

NL = 1

The computations, while easy, can be laborious. However, there is no need to do this by hand. A table can be set up in most spreadsheet programs, such as EXCEL. Many SPC programs include both CUSUM and Shewhart with runs rules as options, as do some statistical packages such as MINITAB.

Sequence	Data	SL	NL	SH	NH	Status
	—	0.000	0	0.000	0	
1	3.38	0.000	0	0.135	1	OK
2	3.32	0.000	0	0.210	2	OK
3	3.13	0.025	1	0.095	3	OK
4	3.25	0.000	0	0.100	4	OK
5	3.22	0.000	0	0.075	5	OK
6	3.09	0.065	1	0.000	0	OK
7	3.13	0.090	2	0.000	0	OK
8	3.16	0.085	3	0.000	0	OK
9	3.30	0.000	0	0.055	1	OK
10	3.29	0.000	0	0.100	2	OK
11	2.99	0.165	1	0.000	0	OK
12	3.18	0.140	2	0.000	0	OK
13	3.14	0.155	3	0.000	0	OK
14	3.13	0.180	4	0.000	0	OK
15	3.05	0.285	5	0.000	0	OK
16	3.11	0.330	6	0.000	0	OK
17	3.13	0.355	7	0.000	0	OK
18	3.21	0.300	8	0.000	0	OK
19	3.00	0.455	9	0.000	0	Off-aim
20	2.97	0.640	10	0.000	0	Off-aim
21	3.16	0.635	11	0.000	0	Off-aim
22	3.06	0.730	12	0.000	0	Off-aim
23	3.15	0.735	13	0.000	0	Off-aim
24	3.14	0.750	14	0.000	0	Off-aim

As you can see, the CUSUM gave an off-aim signal at data point 19. SL = 0.455, which exceeded the design H of 0.36. NL = 9, which indicates that the CUSUM was accumulating deviations lower than the target for the last 9 data entries.

Another feature of CUSUM is the ability to determine the average that caused an out-of-control signal by either SL or SH. If the signal is indicated by SH:

$$\text{Average} = \text{Aim} + K + \frac{\text{SH}}{\text{NL}}$$

If the signal is indicated by SL:

$$\text{Average} = \text{Aim} - \text{K} - \frac{\text{SL}}{\text{NL}}$$

For this example:

$$\text{Average} = 3.20 - 0.045 - \frac{0.455}{9} = 3.10$$

This computation is the same as obtaining the average of the NL = 9 points leading up to the off-aim signal (sequence 11–19). Sometimes there is a tendency to look for assignable causes at the time of the data point causing an off-aim signal. The CUSUM indicates that the best estimate of when the change occurred was 9 data points previous to the off-aim signal (sequence 11). Search for the assignable cause should focus on things that happened about that time.

Following are two control charts of these data from MINITAB:

1. Individuals (I) chart with runs rules similar to AT&T rules (Figure 5.3); and

2. CUSUM chart (Figure 5.4).

An off-aim signal was given at data point 20. Runs rule five was indicated to have caused the signal. Runs rule five is 2 of 3 consecutive points above or below 2 sigma from aim. Points 19 and 20 were 3.00 and 2.97. Both were less than $3.20 - 2.(.09) = 3.02$.

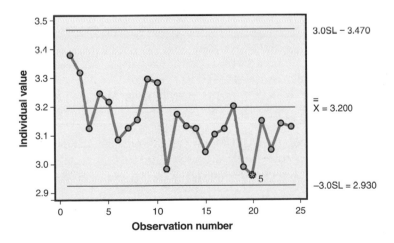

Figure 5.3   I chart for data.

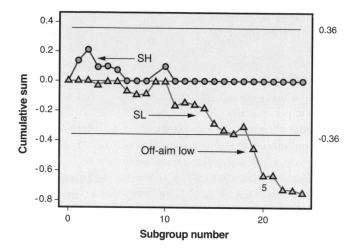

Figure 5.4 CUSUM chart example.

The two lines plotted are SH above the centerline and SL below the centerline. The CUSUM chart shows an off-aim low signal at point 19, where SL is below the lower limit of -0.36.

REFERENCES

1. J. M. Lucas, "A Modified V-Mask Control Scheme," *Technometrics* 15 (1973): 833–847.
2. J. M. Lucas, "The Design and Use of V-Mask Control Schemes," *Journal of Quality Technology* 8 (1976): 1–12.
3. J. M. Lucas, "Combined Shewhart-CUSUM Quality Control Schemes," *Journal of Quality Technology* 14 (1982): 51–59.
4. J. M. Lucas and R. B. Crosier, "Fast Initial Response (FIR) for CUSUM Quality Control Schemes: Give Your CUSUM a Head Start," *Technometrics* 24 (1982): 199–205.

5.4 IDENTIFYING AN OUTLIER

As mentioned in Section 3.6.1, there are many ways of flagging a measurement value as an outlier. One simple yet powerful way is the typical boxplot, a graphical technique. Boxplots could be employed in a quality management system in such a way that every measurement value is compared with previously accepted data, and any outlier flagged as soon after the measurement value has been generated as possible. Another possibility is to do all the data analysis work necessary to come up with a valid estimate of the total standard deviation at one point in time, and then use that standard deviation as the basis for an outlier check. Reestimating this standard deviation periodically, of course, would be required.

If the latter choice is used, the key to flagging a reported measurement result as an outlier lies in the judicious choice of what is called the "outlier standard deviation." One technique used to estimate this standard deviation uses the median absolute deviation (MAD) statistic. There are other statistics that could be used, one being the fourth-spread.[2] MAD is presented here as one good choice for estimating the outlier standard deviation.

MAD is a powerful resistant statistic and is defined as the median of the absolute values of the deviations of each measurement value from the median of all reported measurement values. Resistant means that the statistic is insensitive to small amounts of "misbehaving" data, in particular outliers, for this discussion. The formula is:

MAD = Median [Measurement Result − Median (All Measurement Result)]

Computationally, the median of all the data is calculated first. This median is subtracted from each measurement result to create a set of deviations. The absolute value of the deviations is taken in order to get a set of all positive numbers. Finally, the median of these positive numbers is calculated, and this is the MAD statistic. MAD is very valuable, because when divided by 0.6745, MAD becomes a resistant estimate of the standard deviation for a normally distributed variable.

While MAD does require two passes through the data, both are done without human intervention. Using MAD eliminates the need to remove data by hand (recalculating means and standard deviations) and the time employed to choose the outlier standard deviation is minimized.

So, in order to flag a measurement result as an outlier, we can use:

Outlier Standard Deviation = MAD Standard Deviation

$$= \frac{\{\text{Median} [\text{Measurement Result} - \text{Median} (\text{All Measurement Results})]\}}{0.6745}$$

The "measurement result" is the number reported by the measurement method. This might be an average of some number of measurement values. MAD can be calculated for all kinds of data sets, except one. If more than half the data are the same number, then MAD cannot be used. This should be a very unusual situation, unless the measurement method itself needs to be improved. One can view this caveat as a positive and recommend use of MAD as the basis for an outlier standard deviation. Data sets of size one cannot have a MAD standard deviation estimate; just as such sets cannot have a regular standard deviation estimate.

One question remains: Given an estimate of the outlier standard deviation for a particular normally distributed variable, what do we do to flag a future measurement value or measurement result as an outlier? The answer depends on how much data is involved, how often a data set will be checked for outliers, and how many resources are available to search for the cause of each outlier. In general, we would call a number an outlier if it fell outside the range:

$$\text{Median} \pm k \cdot \text{Outlier Standard Deviation}$$

where: k is a constant to be chosen

If 7–21 individual measurement results are checked for 10 or more properties every week, a value of 5.0 for k is recommended. Over time, of course, as causes of outliers are found and prevention activities implemented, reducing k may be desirable. A value for k much below 4.0 should seldom, if ever, be used. The reason is that too many values that belong to the natural normal distribution will be removed, thereby causing a standard deviation that is actually too small. Future values would be tested using this small standard deviation, and still more valid values would be removed. So, there is grave danger in using any rule that blindly remeasures or resamples based on a value being more than some arbitrary number of standard deviations away from the median.

REFERENCES

1. M. G. Natrella, *Experimental Statistics—Handbook 91* (Washington D.C.: National Bureau of Standards, 1963).
2. C. H. Hoaglin, F. Mosteller, and J. W. Tukey, eds., *Understanding Robust and Exploratory Data Analysis* (New York: John Wiley & Sons, 1983).

5.5 RUGGEDNESS TEST EXAMPLE

As mentioned in Section 2.2.8, "The Need for Ruggedness Tests," ruggedness tests help identify test variables which most strongly influence test results.[1] Application: Gas chromatographic trace measurement of a residue using a flame photometric detector (phosphorus mode).

The measurement method had been optimized. However, we were interested in ruggedness of the measurement method—how much minor departures from the measurement method would change the measurement values. We chose an N = 8 design (Table 5.1), and selected seven variables which we varied within reasonable levels (Table 5.2). Eight samples of a solution originally measured at 4.92 ppb residue were analyzed and treated as indicated in Table 5.1 and Table 5.2; that is, sample 1 had the treatment levels A,B,C,D,E,F,G; sample 2 had A,B,c,D,c,f,g, and so on.

The measurement value of each sample is designated by a lower case letter, s through z (Table 5.3). To evaluate the effect of changing factor *A* to *a*, the average of *A* treatments, (s+t+u+v)/4, was compared to the average of *a* treatments, (w+x+y+z)/4. The same was done for the other pairs: *B* (s+t+w+x)4 and *b* (u+v+y+z)/4, and so on (Table 5.4). The absolute value of the differences was calculated, and the samples ranked in order of descending difference (Table 5.5).

Table 5.1 N = 8 design.

Sample number							
1	2	3	4	5	6	7	8
Factor combination							
A	A	A	A	a	a	a	a
D	B	b	b	B	B	b	b
C	c	C	c	C	c	C	c
D	D	d	d	d	d	D	D
E	e	E	e	e	E	e	E
F	f	f	F	F	f	f	F
G	g	g	G	g	G	G	g
Results							
s	t	u	v	w	x	y	z

Table 5.2 Variable levels for ruggedness test.

Variable	Level	
Column temperature, °C	A = 175	a = 185
Hydrogen flow, ml/min	B = 180	b = 200
Air flow, ml/min	C = 60	c = 50
Oxygen flow, ml/min	D = 20	d = 25
Injector temperature, °C	E = 200	e = 250
Carrier flow, ml/min	F = 60	f = 70
Injection size, 1 µL	G = 2	g = 3

Table 5.3 Measurement values.

Sample number	Area, corrected for injection size	Residue, ppb	Result
1	3.172	4.321	s
2	3.647	4.969	t
3	3.703	5.045	u
4	3.647	4.969	v
5	3.957	5.391	w
6	3.358	4.575	x
7	3.801	5.178	y
8	3.580	4.877	z

Table 5.4 Ruggedness test calculations.

	Upper case letter	Lower case letter	Difference
A-a	19.304/4 = 4.826	20.021/4 = 5.005	-0.179
B-b	19.256/4 = 4.814	20.069/4 = 5.017	-0.203
C-c	19.935/4 = 4.984	19.390/4 = 4.848	0.136
D-d	19.345/4 = 4.836	19.980/4 = 4.995	-0.159
E-e	18.818/4 = 4.705	20.507/4 = 5.127	-0.422
F-f	19.558/4 = 4.890	19.767/4 = 4.942	-0.052
G-g	19.043/4 = 4.761	20.282/4 = 5.071	-0.310

Std. Dev. = s = 0.336

Table 5.5 Results of ruggedness test.

Ranking of differences (Absolute values)	
Variable	**Average difference, ppb residue**
Injection port temperature	0.422
Injection size	0.310
Hydrogen flow	0.203
Column temperature	0.179
Oxygen flow	0.159
Air flow	0.136
Carrier flow	0.052

The ranking shows that the injection port temperature had the most effect, followed by injection size. This is not surprising, since we knew that the residue degrades in the injection port, and the measurement method actually measures the degradation product.

REFERENCES

1. W. J. Youden, *Statistical Techniques for Collaborative Tests* (Gaithersburg, MD: International Association of Official Analytical Chemists, 1967): 31.
2. F. Yates, "Complex Experiments," *J. Roy. Statistical Society (Supplement)* 2 (1935): 181–247.
3. W. J. Youden, "Design for Multifactor Experimentation," *Industrial and Engineering Chemistry* 51 (1959): 70A–80A.
4. ASTM, E1169-89, *Standard Guide for Conducting Ruggedness Tests* (West Conshohocken, PA: American Society for Testing of Materials, 1996).
5. AOAC International, *Intralaboratory Analytical Method Validation* (Gaithersburg MD: International Association of Official Analytical Chemists, 1977): 10-13–10-15.

5.6 A NEW MEASURE FOR MEASUREMENT SYSTEM IMPROVEMENT EFFORTS

As mentioned in Section 3.5, "Controlling the Measurement System," if maintenance sampling is in place, a new continual improvement measure becomes available that does not have the deficiencies associated with measures based on observed data, such as capability and performance indices. If either VST (short-term measurement variability) or VLT (long-term measurement variability) is too large and a reduction in measurement variability is deemed necessary, for example, where VST and VLT come from maintenance sampling data, then the ILC variance components can be a primary source of ideas concerning what to do. No additional data collection may be necessary. This is a useful statistic to use for a measurement system improvement measure.

Measurement System Improvement Measure (MIM) = $\sqrt{(VMEAS)}$

where:

$$VMEAS = VST + VLT$$

Since this measure (MIM) is actually a standard deviation, it directly measures the measurement system variability. If an index is desired, dividing the current MIM by the initial MIM is the first properly determined MIM:

Measurement System Improvement Index (MII) =

$$\frac{\sqrt{(VMEAS)}}{[\text{Initial } \sqrt{VMEAS}]} = \frac{MIM}{(\text{Initial MIM})}.$$

This index essentially provides the current measurement system standard deviation as a percent of the initial measurement system standard deviation, thereby giving immediately usable information on the current state of a measurement system.

5.7 CONTINUALLY UPDATING ESTIMATES OF MEASUREMENT SYSTEM VARIANCE COMPONENTS

The following describes a procedure that provides ongoing estimates of measurement variability and product variability. This is called maintenance sampling. The procedure is discussed in more detail in Juran's *Quality Handbook*, Fifth Edition.[1]

The procedure defines the total variability consisting of five variances.

$$V(total) = VLL + VPP + VWP + VLT + VST$$

where:

VLL is lot-to-lot variability

VPP is product unit to product unit variability

VWP is within-product unit variability

VLT is long-term measurement variability

VST is short-term measurement variability

The variance of any average can be stated as:

$$\text{Variance} = \frac{VLL}{NLL} + \frac{VPP}{NPP} + \frac{VWP}{NWP} + \frac{VLT}{NLT} + \frac{VST}{NST}$$

Where:

VST is the variability associated with the measurement at one point in time. It includes the following sources of variation:

- Measurement method

- Sampling procedure

- Sample preparation

- Calibration (when performed before each measurement)

- Within-sample variability

VLT is additional variability associated with a measurement over and above the variability at a given time. It includes such sources as:

- Calibration (when performed periodically or on demand)
- Multiple measurement equipment
- Sampling procedure
- Reagents
- Operators (sampling)
- Operators (testing)
- Ambient conditions

VLT is extremely important because any deviation causes a bias affecting *all* measurements at a given time. It is not usually considered because people either do not think about it, do not know how to handle it, or do not consider the consequences of it.

VWP is the variability within a product unit. It is sometimes called sample-sample variability.

VPP is the variability of product units within a lot.

VLL is the lot-to-lot variability.

NLL is the number of lots included in the average.

NPP is the number of product units (sublots) included in the average.

NWP is the number of within-product units (increments) included in the average.

NLT is the number of submissions (or measurement times) included in the average.

NST is the number of individual measurements in the average.

For example, if you want to know the variance of any given lot average consisting of a composite of four product units measured in triplicate at one measurement time, the variance is:

$$\text{Variance} = \frac{\text{VPP}}{4} + \frac{\text{VWP}}{4} + \frac{\text{VLT}}{1} + \frac{\text{VST}}{12}$$

The basic procedure for estimating these components is the ABCD contrast method. In this example, it was assumed that two or more product units (sublots) were sampled from each product lot. Once per day an additional ABCD sample was taken. The routine product samples were used to estimate lot-to-lot and within-lot variance. The ABCD samples were used to estimate VWP, VLT, and VST; by combining the ABCD estimates with the routine estimates, estimates of all five variance components can be obtained.

The following describes ABCD sampling. Once per day:

- Select one product unit (sublot) from one lot and take two large samples from different locations in the product unit (sublot).

- Split each sample into two smaller samples. Do not blend or otherwise attempt to make them uniform. Identify the samples from the first location as A and B and from the other location as C and D.

Submit samples A and C "today."
Submit samples B and D at another time, "tomorrow" or "next shift."
The sample pattern is:

		Submission Time	
		1	2
Within Product Unit	1	A	B
	2	C	D

Four contrasts are computed for each ABCD sample set:

$$C1 = LL \text{ and } PP = \frac{(+A + B + C + D)}{4}$$

$$C2 = WP = \frac{(-A - B + C + D)}{2}$$

$$C3 = LT = \frac{(+A - B + C - D)}{2}$$

$$C4 = ST = \frac{(+A - B - C + D)}{2}$$

After a minimum of 35 ABCD sets (preferably 60 or more), compute the variance of the 35 or more contrasts for each contrast type.

The expected means squares are:

$$V(C4) = VST$$

$$V(C3) = VST + 2VLT$$

$$V(C2) = VST + 2VWP$$

$$V(C1) = VST + \frac{(VLL + VPP)}{4}$$

VST, VLT, and VWP can be computed from these relationships.

$$VST = V(C4)$$

$$VLT = \frac{[V(C3) - V(C4)]}{2}$$

$$VWP = \frac{[V(C2) - V(C4)]}{2}$$

ANOVA of the data yields the lot-to-lot variance and within-lot variance:

The lot-to-lot variance is an estimate of VLL + VLT

The within-lot variance is an estimate of VPP + VWP + VST

Thus, VLL and VPP can be determined as soon as the ABCD analysis is done.

There are several modifications to the ABCD plan. Some modifications allow estimates of all five variance components from the contrast data, others are aimed at minimizing the sampling. The spreadsheet analysis example on the following page does not contain enough data for reliable estimates of the variance components (n should be 60 to 90).

An additional feature allows a check to see if there is a bias in any of the contrasts. For example, if the C3 or LT contrast shows a significant bias, it may indicate a change in the product from one time to another, such as aging. An important plus is that all contrast variances have the same degrees of freedom showing an efficient use of the data.

REFERENCE

1. J. M. Juran, *Quality Handbook*, Fifth edition (New York: McGraw-Hill, 1998): chapter 27.

ABCD DATA

Day	A	B	C	D
1	124	119	119	118
2	110	107	113	116
3	106	108	110	108
4	98	100	99	96
5	101	100	99	94
6	87	84	86	80
7	106	105	106	104
8	115	112	118	115
9	101	97	100	102
10	107	108	108	109
11	107	106	110	110
12	102	102	101	102
13	89	89	86	89
14	113	112	117	118
15	104	105	101	99
16	114	111	111	110
17	103	101	98	95
18	121	122	122	122
19	95	90	92	90
20	122	122	120	120
21	124	125	123	123
22	105	100	99	106
23	102	105	102	102
24	144	141	139	140
25	103	106	106	105

ABCD CONTRASTS

	AVG	WP	LT	ST
Day	C1	C2	C3	C4
1	120.00	-3.0	-3.0	2.0
2	111.50	6.0	0.0	3.0
3	108.00	2.0	0.0	-2.0
4	98.25	-1.5	-0.5	-2.5
5	98.50	-4.0	-3.0	-2.0
6	84.25	-2.5	-4.5	-1.5
7	105.25	-0.5	-1.5	-0.5
8	115.00	3.0	-3.0	0.0
9	100.00	2.0	-1.0	3.0
10	108.00	1.0	1.0	0.0
11	108.25	3.5	-0.5	0.5
12	101.75	-0.5	0.5	0.5
13	88.25	-1.5	1.5	1.5
14	115.00	5.0	0.0	1.0
15	102.25	-4.5	-0.5	-1.5
16	111.50	-2.0	-2.0	1.0
17	99.25	-5.5	-2.5	-0.5
18	121.75	0.5	0.5	-0.5
19	91.75	-1.5	-3.5	1.5
20	121.00	-2.0	0.0	0.0
21	123.75	-1.5	0.5	-0.5
22	102.50	0.0	1.0	6.0
23	102.75	-1.5	1.5	-1.5
24	141.00	-3.0	-1.0	2.0
25	105.00	1.0	1.0	-2.0

ABCD CONTRASTS SUMMARY

Contrast	Mean	Variance(V)
C1	107.380	151.7767
C2	-0.440	8.2567
C3	-0.760	2.8567
C4	0.280	3.8767

VARIANCE COMPONENTS

		(Contrast Variances)
VLL+VPP	149.9675	V1 + (V4 - V2 - V3)/4
VWP	2.1900	(V2 - 4)/2
VLT	-0.5100	(V3 - 4)/2
VST	3.8767	V4

Significant Bias? $t_{.05} = 2.064$ df = 24

Contrast	Mean	95% CL	t
C2	-0.440	1.186	-0.766
C3	-0.760	0.698	-2.248***
C4	0.280	0.813	0.711

6

Glossary

Unless otherwise indicated, the definitions in this glossary are those of the authors.

Accuracy—(a) The closeness of agreement between a single measurement value and the accepted reference material measurement value reported by the reference laboratory. Note: The term *accuracy*, when applied to a set of measurement values, involves a combination of random components and a common systematic component. (b) The difference of individual measurement values from the "true" or "assigned" or "accepted" measurement value.

ARL—Average run length—the expected number of subgroups necessary to detect a given shift in the average.

Bias—The difference between the measurement value obtained from the measurement system under consideration and an accepted reference laboratory measurement value. Note: Bias is the total systematic variance as contrasted to random variance. There may be one or more systematic variance components contributing to the bias. A larger systematic difference from the accepted reference laboratory measurement value is reflected by a larger bias number.

Calibration—The statistical comparison of the measurement results from a standards reference laboratory's measurement system to those of the measurement system under evaluation. *See* Section 5.1 and Section 5.2.

Confidence Interval—The end points of the interval about a sample statistic that is believed, with a specified confidence coefficient, to include the population parameter.

Contrast—Algebraically, a contrast is a mathematically defined difference describing a comparison of interest. For example, in a two level factorial experiment, the factor effects are determined by contrasts. The effect of X1 is defined by the "contrast."

X1	X2	Y
-1	-1	20
+1	-1	25
-1	+1	30
+1	+1	35

$$\frac{(-Y1 + Y2 - Y3 + Y4)}{2}$$

$$\frac{(-20 + 25 - 30 + 35)}{2} = \frac{10}{2} = 5$$

This is a comparison of the data associated with X1 = -1 and the data associated with X1 = +1.

Corrective Action—To search for and eliminate the root causes of nonconformances that have occurred. Repair is not corrective action.

C&PI—Chemical and Process Industries.

Degrees of Freedom—The number of independent comparisons available to estimate a parameter (nu or $v = n - 1$).

Hypothesis—A tentative assumption made in order to draw out and test its logical or empirical consequences.

Maintenance Sample—The portion of a routine product sample that is retained for later repeat measurements. A maintenance sample is an instance of control material (CM). The retained product samples must be homogeneous and stable. Packaging, storage, and handling must not change the retained sample's properties.

Maintenance Sampling Data—The measurement values and measurement results from a maintenance sampling program.

Maintenance Sampling Program—The systematic submission of retained product samples (CM) for repeat measurements. The submissions are done at time intervals selected to permit the estimation of short- and long-term variation of the product and the measurement system. The stability of the material may have a major impact on the interval selected and on the meaning of the measurement results over time.

Measurement Method—A qualitative or quantitative comparison of two things at a time. This comparison generates a measurement value. Measurement method as used in this book is equivalent to the terms "test method" or "test procedure."

Measurement Result—The number derived from a measurement value or series of measurement values. The measurement result can be a simple mathematical manipulation of a measurement value; for example, the conversion of a pychnometer measurement value into a density measurement result, or 95 percent active ingredient calculated from 20 ml of titrant, or the calculation of an average of several measurement values. *See* Result.

Measurement System—The measurement system starts with taking the sample to be measured from the population through generation of the measurement value to delivering the measurement result needed to make the decision. This system is often called the measurement process.

Measurement Value—The number read from a measurement device or instrument. For example, the 20 milliliters read from a titration burette is a measurement value.

Parameter—In statistical terms, a constant or coefficient that describes some characteristic of a population (for example, average [μ], standard deviation [σ], variance [σ^2]). In process engineering, parameter is often used to define a condition of the process that can be set to a specified value. The term is often interchangeable with the term *variable*.

Population—The totality of items or units of material under consideration.

Precision—A generic concept related to the closeness of agreement between randomly selected measurement values obtained under stipulated conditions.

Preventive Action—To search for and eliminate root causes that have the potential of causing a nonconformance before a nonconformance occurs.

Repeatability—The variability in the measurement values generated at the same time, with the same equipment, by the same operator, following the same measurement method, on the same sample.

Repeatability Conditions—Conditions where independent measurement values are obtained with the same measurement method, on identical samples, in the same laboratory, by the same operator, using the same equipment, within short intervals of time. Note: The same operator means that for a particular step in a measurement method, one person performs that step for all samples.

Representative Sample—A portion of a population that is expected to exhibit the same properties in the same proportions as the population. For example, a portion of material taken from a larger bulk container or a duration of time during the total time of operation of a process.

Reproducibility—The number below which the absolute difference between single measurement values on identical material obtained by operators in possibly different laboratories, using the agreed measurement method, may be expected to lie within a specific probability (typically 95% and hence R = 2.8 S_R). This is precision between measurement systems.

Reproducibility Conditions—Conditions where measurement values are obtained using the same measurement method, on identical samples, in possibly different measurement laboratories, with different operators, using different equipment.

Result—The reader will note that the use of the word *result* in this book without a modifier refers to the answer or outcome from a calculation involving a mathematical function or equation, or the application of a statistical method or technique. *See* Measurement Result.

Sample—A portion or a single item taken from a larger whole or group (population).

Specimen—The portion of a material that is the actual input to the measurement method. Specimens in the C&PI are often destroyed during testing, preventing multiple measurement values being generated on the exact same specimen.

Statistic—A number calculated from a sample of observations, most often to form an estimate of a population parameter (for example, $\overline{X}$ or s to estimate μ or σ).

UOM—Unit of measure is the incremental amount differentiated by the measurement scale.

Value—A point on a scale. Most commonly a number such as the amount of material in grams, or the length of an object in centimeters.

Variable—In statistical terms, *variable* is a condition that one has control over and that can be set to different specified values for experimental purposes. In process engineering, the term is interchangeable with *parameter*.

REFERENCE

1. Ronald E. Walpole, Raymand H. Myers, and Sharon L. Meyers, *Probability and Statistics for Scientists and Engineers* (New Jersey: Prentice Hall, 1998): 648.

Index